Transactions on Computer Systems and Networks

Transactions on Computer Systems and Networks is a unique series that aims to capture advances in evolution of computer hardware and software systems and progress in computer networks. Computing Systems in present world span from miniature IoT nodes and embedded computing systems to large-scale cloud infrastructures, which necessitates developing systems architecture, storage infrastructure and process management to work at various scales. Present day networking technologies provide pervasive global coverage on a scale and enable multitude of transformative technologies. The new landscape of computing comprises of self-aware autonomous systems, which are built upon a software-hardware collaborative framework. These systems are designed to execute critical and non-critical tasks involving a variety of processing resources like multi-core CPUs, reconfigurable hardware, GPUs and TPUs which are managed through virtualisation, real-time process management and fault-tolerance. While AI, Machine Learning and Deep Learning tasks are predominantly increasing in the application space the computing system research aim towards efficient means of data processing, memory management, real-time task scheduling, scalable, secured and energy aware computing. The paradigm of computer networks also extends it support to this evolving application scenario through various advanced protocols, architectures and services. This series aims to present leading works on advances in theory, design, behaviour and applications in computing systems and networks. The Series accepts research monographs, introductory and advanced textbooks, professional books, reference works, and select conference proceedings.

Tanvir Mustafy · Md. Tauhid Ur Rahman

Statistics and Data Analysis for Engineers and Scientists

 Springer

Tanvir Mustafy
Dhaka, Bangladesh

Md. Tauhid Ur Rahman
Dhaka, Bangladesh

ISSN 2730-7484   ISSN 2730-7492 (electronic)
Transactions on Computer Systems and Networks
ISBN 978-981-99-4660-0   ISBN 978-981-99-4661-7 (eBook)
https://doi.org/10.1007/978-981-99-4661-7

This Springer imprint is published by the registered company Springer Nature Singapore Pte Ltd.
The registered company address is: 152 Beach Road, #21-01/04 Gateway East, Singapore 189721,
Singapore

Paper in this product is recyclable.

Contents

About the Authors

Dr. Tanvir Mustafy, SEng, EIT received his B.Sc. degree in Civil Engineering from BUET, Bangladesh, in 2011, an M.Sc. degree in structural engineering from the University of Alberta, Canada, in 2013, and a Ph.D. Degree in Computational Mechanics from University of Montreal, Canada, in 2019. His research and teaching interests include the theory and application of machine learning, structural engineering, earthquake engineering, advanced finite element modeling, dynamics of structures, data analysis, and injury biomechanics. Dr. Mustafy currently serves as an Associate Professor in the Department of Civil Engineering at the Military Institute of Science and Technology (MIST), Bangladesh. Before joining MIST, Dr. Mustafy worked as a member of a prestigious scientist group led by one of the most renowned Research Chairs in Canada during his doctoral period. He traveled to France as a visiting scholar and spent three months working at Aix-Marseille University. Dr. Mustafy also received the prestigious Professional Structural Engineer (SEng.) by BUET, ICC (USA), RAJUK, URP and IEB in 2023.

Dr. Md. Tauhid Ur Rahman finished his Ph.D. in Environmental Engineering from Tohoku University, Japan, in 2009. He did his M.Sc. in Environmental Engineering, Land, and Water Engineering from KTH, Sweden, and his B.Sc. in Civil Engineering from Bangladesh University of Engineering and Technology, Bangladesh. He is currently working as a professor in the CE Department of MIST. His research interests are water quality modeling, land use change detection, climate change, water insecurity, micro-climate effect, etc.

Chapter 1
Introduction

Abstract The inaugural chapter of "Statistics and Data Analysis for Engineers and Scientists" serves as the cornerstone for a comprehensive journey into the world of statistics, offering a solid introduction to fundamental concepts and their practical applications in the realms of engineering and science. Starting with a broad overview, the chapter underscores the pivotal role that statistics plays in deciphering data and guiding informed decision-making. It emphasizes the intrinsic importance of scientific data within engineering domains, highlighting how statistics acts as a potent tool for extracting valuable insights. This chapter introduces two bedrock concepts: population and sample, essential for understanding statistical principles. It delineates variables of interest, statistical populations, and samples, laying the groundwork for discussions on data collection and analysis. Addressing the inherent variability in scientific data, the chapter underscores the need for robust statistical methodologies to navigate this variability effectively. The chapter then delves into data collection, with an emphasis on experimental design and the rationale behind random assignment of experimental units. It instills in readers the importance of sound data collection practices. Measures of location, including the sample mean and median, are introduced to facilitate comprehension of central tendencies within data. The chapter also covers additional measures of location, providing a comprehensive perspective on data summarization. Further exploration unfolds in measures of variability, encompassing the sample standard deviation and sample range. Clarification on units of standard deviation and variance aids in grasping nuances related to data dispersion. In preparation for subsequent chapters, the chapter draws a distinction between discrete and continuous data. The chapter culminates with an introduction to statistical modeling, scientific inspection, and graphical diagnostics through visual tools like scatter plots, stem-and-leaf plots, histograms, box-and-whisker plots, and other distinguishing features of a sample. To solidify comprehension, the chapter offers a range of exercises aimed at reinforcing key concepts, fostering active learning, and equipping readers with essential skills for their continued exploration of statistical methodologies throughout the book. In essence, this first chapter provides a foundational framework for readers, arming them with the requisite knowledge to embark on a comprehensive exploration of statistics and data analysis within the context of engineering and science.

T. Mustafy and Md. T. U. Rahman, *Statistics and Data Analysis for Engineers and Scientists*, Transactions on Computer Systems and Networks,
https://doi.org/10.1007/978-981-99-4661-7_1

Keywords General overview · Use of scientific data · Importance of statistics · Statistics and engineering · Population and sample · Measures of location · Measures of variability · Experimental design · Random assignment · Graphical diagnostics · Variable of interest · Variability in scientific data · Discrete and continuous data · Statistical modeling · Box-and-whisker plot

1.1 General Overview

Fundamentally, statistics are used to know occurrences or non-occurrences of certain events and draw valid inferences or conclusions. A successful business can make decisions quickly and accurately. It understands what the consumers want. Thus, it should be able to determine what to manufacture and sell and what amounts. Since the 1980s, much effort has been placed into improving business functioning. This is where knowledge of applied statistics comes in handy. The application of statistical tools and statistical thinking among management employees has been credited with economic success in many countries.

1.1.1 Use of Scientific Data

Statistics may be pretty helpful when a considerable amount of data is not organized. When a corporation employs statistics to obtain insights, it makes a complex process appear simple.

Scientific data is required to apply statistical methods in industrial production, engineering, food technology, computer software, power sources, medicines, and other fields. Data is gathered, summarized, reported, and archived for future reference. However, there is a significant difference between scientific data collection and inferential statistics. In recent decades, the latter has received broad attention (Fig. 1.1).

Statistical techniques are used to analyze data from a process to determine where improvements to process quality may be made. During the decision-making process, the inherent variability in research must be considered. For example, in medical statistics, variation among patients is commonly seen even for the same illness.

In a biological trial of a new medicine that lowers blood pressure, 87% of patients reported relief. However, it is widely accepted that the present treatment, or "old" drug, relieves 80% of chronic patients with hypertension. However, the new medicine is more expensive to produce and may have unwanted side effects. Should the new medication be used? This is a dilemma that pharmaceutical companies routinely face in collaboration with the Federal Drug Administration (FDA). Variability must be taken into account once again. The "87%" figure is based on a certain number of patients chosen for the research. Perhaps, the observed number of "successes" would be 70% if the research were redone with additional patients.

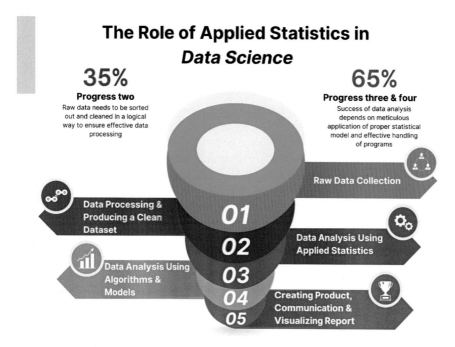

Fig. 1.1 Role of applied statistics in data science

1.1.2 Importance of Statistics

When analysts appropriately use statistical processes, they provide more reliable findings. Statistical analyses account for the results' uncertainty and mistake. These are some of the steps used in statistical research:

- Collecting data that can be trusted.
- Analyzing the data in the right way.
- Drawing reasonable conclusions.

The answers produced by statistical analysis can be used to make better judgments and choices. For example, local officials would want to know if iron levels in the residential water supply are within acceptable limits. Because not all water can be tested, the survey must be based on only a portion of the data obtained from water samples gathered. A mechanical engineer, for example, may need to determine the strength of generator supports in a power plant. He gains the strengths of a few supports by loading them to failure. These values serve as a benchmark for evaluating the strength of any other untested supports.

When data is needed, statistical theories recommend a four-step gathering approach.

i. Clearly state the investigation's objectives.

ii. Make a list of what data is needed and how to obtain it.
iii. Use appropriate statistical methods to extract information from the data efficiently.
iv. Analyze the data and draw appropriate conclusions.

These procedures will serve as a frame of reference as we build the fundamental concepts of statistics. Statistical approaches and reasoning can help us more efficiently gather data and draw valuable conclusions.

When drawing a statistical inference or a conclusion that goes beyond the information contained in a dataset, always proceed with caution. Thinking about how far we can extrapolate from an information set is essential. It is crucial to consider if such broad generalizations are suitable or warranted and whether further data is needed. We treat statistics as a science wherever possible, building each statistical theory from its probabilistic foundation and applying each notion to physical or engineering scientific problems as soon as it is developed. The method in which statistical inferences are drawn about known but fixed variables will be the basis for the ways we address these challenges.

1.1.3 Statistics and Engineering

Statistical approaches may be helpful in engineering to create new products and systems, refine current designs, and invent and improve manufacturing processes. In engineering and industrial management, the recent expansion of statistics has had a significant influence. Indeed, it is impossible to exaggerate the importance of statistics in addressing industrial issues, maximizing the efficient use of resources and labor, conducting fundamental research, and developing new goods.

It plays a function in a variety of engineering domains:

- Statistics are used to design experiments to test and create models of engineering components and systems.
- Statistics aid in quality control in the manufacturing sector.
- Reliability engineering uses statistics to assess a system's capacity to serve its intended purpose and methods for enhancing its performance.
- Probability and statistics are applied in product and system design.

Engineers working on designing and manufacturing new goods must adhere to a statistical strategy for anticipating and resolving quality issues before production. Statistics, like other disciplines, has become an indispensable tool for engineers. It helps people comprehend events prone to change and forecast or govern them successfully. Our focus will be on engineering applications, but we will not hesitate to refer to other domains to demonstrate to the reader the broad applicability of most statistical approaches.

1.1.4   Two Basic Concepts—Population and Sample

Developing a clear, well-defined statement of purpose is the first stage in any study. Data that is relevant must be gathered. However, from a practical standpoint, it is sometimes difficult or impractical to gather whole datasets. No matter how much experimentation is done using data collected from laboratory research, there is always more that might be done.

Understanding statistics requires distinguishing between the collected data and the vast amount of data that is not considered. A unit is the basis of each measurement. A single object or a person whose features are of interest is referred to as a unit. The total number of units for which information is requested is referred to as the population of units.

When the population is large, geographically dispersed, or difficult to contact, it is necessary to use a sample. With statistical analysis, sample data may be used to estimate or test population data hypotheses. A statistical population collects all measurements of each unit in the entire population of units for whom data is available. A subset of observations obtained in a study is referred to as a sample from a statistical population. Selecting a sample is the first step in any statistical approach to learning about the population. The sample must be representative of the general population and large enough to include enough information to answer the critical questions about the demographic (Fig. 1.2).

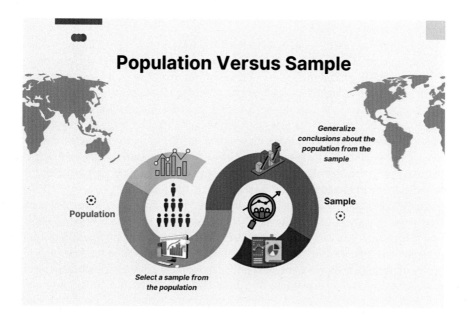

Fig. 1.2 Population versus sample

1.1.5   Variable of Interest, Statistical Population, and Sample

• Why use samples?
Because of its size or inaccessibility, it is sometimes impossible to research the entire population
Collecting data from a sample is easier and more efficient
Fewer expenses are associated with participants, laboratories, equipment, and researchers when working with a sample
It is easier to execute statistical analyses on smaller datasets, making them more manageable

1.1.6   Variability in Scientific Data

Statistical approaches to solving the difficulties outlined above require dealing with variability. For example, there would be no need for statistical approaches in a manufacturing company if the observed product density was always the same. No statistical analysis would be required if an instrument for detecting carbon dioxide consistently produced the same value and the value was always accurate. Life would be simple for scientists at pharmaceutical companies if there were no patient-to-patient variability in medication. Without variability in data, no statistician would be required in the decision-making process.

Populations are used to gather samples. A population can sometimes represent a scientific system. In a drug trial, a group of patients may be selected and offered the new medicine. Based on the results obtained from this group, conclusions may be drawn regarding the effectiveness of the new medicine for the entire population. Here, the selected group is the sample representing the entire population. It is critical to collect scientific data methodically, with "planning" at the top of the priority list. For example, a manufacturing company may want to inspect variation in product density. An engineer would need to investigate the impact of temperature, pH, humidity, catalyst, and other variables. They can systematically adjust these variables to whatever amounts are recommended by the prescription or study design.

On the other hand, a forest ecologist interested in studying the factors that determine wood thickness in a particular tree species cannot necessarily construct an experiment. In this instance, observational research may be required. Both of these investigations lend themselves to statistical inference methods. In the first case, the quality of the conclusions will be determined by how well the experiment is planned. In the latter, the scientist is limited by the information available. It is unfortunate, for example, if an agriculturist wants to research the influence of rainfall on crop cultivation and the data is collected during a drought.

The use of statistical thinking by managers and statistical reasoning by scientists is well-known. For researchers, scientific data is precious. A researcher may

Table 1.1 Stem weights (g) after 120 days

No phosphorus	0.52	0.5	0.37	0.47	0.35	0.45	0.28	0.43	0.35	0.33
Phosphorus	0.47	0.33	0.65	0.64	0.43	0.82	0.64	0.87	0.56	0.38

only require a summary of the data included in a sample. In other words, inferential statistics are not always required. It is better to use a set of single-number
statistics or descriptive statistics. These statistics indicate the general nature of the
sample's distribution and variability. With descriptive statistics, graphics are sometimes provided. With modern statistical software programs, single-number statistics such as percentiles and standard deviations alongside graphs representing the
sample's nature are all possible.

Example 1.1

- **Problem Statement**: The following data is from research on establishing a link
  between tree roots and fungal activity. Two groups of ten red oak seedlings were
  planted in a greenhouse. One group contained phosphorus-treated seedlings while
  the other did not. The rest of the surroundings remained unchanged. A particular
  species of fungus was found on the seedlings. After 120 days, the stem weights
  in grams were measured, as shown in Table 1.1. Two samples from two different
  populations are used in this example. How can we determine whether adding
  phosphorus to the soil affects root development?
- **Solution**: If we plot the data in Table 1.1, we can identify a trend. The line
  graphs show that using phosphorus increases stem weight on average. The phosphorus observations are noticeably higher than the non-phosphorus observations.
  It suggests that phosphorus is helpful, but how can we put a number on it? How
  can all of the visible evidence be summed up in some way? The foundations of
  probability may be used, just like in the previous case. More quantitative methods
  to summarize data will be discussed in the upcoming sections (Fig. 1.3).

1.2 Collection of Data

Data collection is the first step for any type of statistical analysis. The technique of
data collecting has a substantial impact on the final outcome of the research. It is
crucial that data is collected using fair or unbiased means.

1.2.1 Experimental Design

In experimental design, the concept of randomness or random assignment is fundamental. However, it is helpful to offer a quick explanation in the context of random

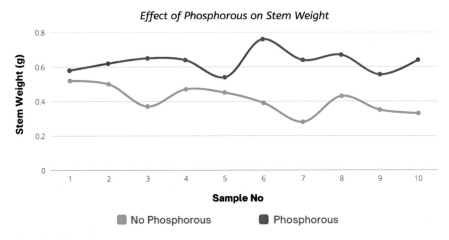

Fig. 1.3 Effect of phosphorous on stem weight

sampling. The populations to be examined or compared comprise a set of treatments or combinations. The phosphorus versus no phosphorus treatments in Example 1.1 illustrate this concept. Many studies include both a control and an experimental group; some researchers reserve the word "experiment" for study designs that include a control group.

Example 1.1 shows that the experiment involves a total of 20 diseased seedlings. The seedlings are distinct from one another, as evidenced by the statistics. There is a lot of variation in stem masses within the phosphorus group (or the no-phosphorus group). This variation is caused by what is known as the experimental unit, which is a crucial topic in inferential statistics that will not be covered in this chapter. It is vital to consider the variability's nature. If there is a great deal of variability, owing to excessive non-homogeneity in experimental units, the variability will "wash out" any discernible difference between the two populations. Remember that this did not occur in this circumstance.

1.2.2   Why Assign Experimental Units Randomly

What are the consequences of not assigning experimental units to treatment combinations at random? It may be particularly evident in the instance of a drug trial. Age, sex, and weight are patient factors that cause variation in the outcomes. Assume that the placebo group, by chance, has a sample of people disproportionately heavier than those in the treatment group. Those who are overweight may have higher blood pressure. As a result, any conclusion reached through statistical inference might have less to do with the medicine and more to do with weight disparities between the two patient samples.

The significance attached to the term variability should be highlighted. Scientific discoveries are "covered up" by excessive heterogeneity among experimental units. In following sections, we shall attempt to characterize and quantify measures of variability. Particular quantities may be computed to convey the nature of the sample in terms of the data's center of placement and variability. These measurements, which assist in describing the dataset's characteristics, are classified as descriptive statistics.

Descriptive statistics differ from inferential statistics. It summarizes a sample rather than trying to learn about the population that the sample represents. Unlike inferential statistics, descriptive statistics are typically nonparametric and do not require formal logic.

Example 1.2

- **Problem Statement**: Suppose we conduct a study to see if covering iron with a corrosion retardant reduces the amount of corrosion. The coating is a protective layer designed to reduce fatigue damage in this material. The effect of humidity on the quantity of corrosion is also of interest. "Thousands of cycles to failure" is used to express a corrosion measurement. Two coating conditions are used: no coating and chemical corrosion coating. Furthermore, there are two relative humidity levels: 20% relative humidity and 80% relative humidity.

  The experiment includes four different treatment options mentioned in Table 1.2. Eight experimental units are in total, all iron specimens; two are randomly assigned to each treatment combination. The result is based on the average of two specimens. What inferences can be made from the given data?

- **Solution**: The following points may be inferred from the data:
  - Fewer cycles remain till failure for samples at higher humidity levels, which indicates that humidity increases corrosion.
  - Coated samples have much higher corrosion resistance than non-coated samples.
  - The designer methodically chose the four treatment combinations. However, variability measures should be taken into account when analyzing the samples.

Table 1.2 Dataset for Example 1.2

Coating type	Humidity levels (%)	Average corrosion in thousands of cycles to failure
Uncoated	20	890
	80	430
Coated	20	1560
	80	1320

1.3   Measures of Location: The Sample Mean and Median

Measures of location are intended to give the analyst some numerical values for where the data's center, or another place, is located. In Example 1.2, the phosphorous sample's central value seems notably greater than the no-phosphorous samples.

Mean is the most commonly used measure of central tendency. It represents the average of the given collection of data. It is applicable for both continuous and discrete data. It is equal to the sum of all the values in the sample divided by the total number of values. The sample mean is an obvious and useful metric. It is utilized in practically every academic subject, such as economics, archeology, sociology, mathematics, and statistics. Per capita income, for example, is the arithmetic average income of a country's population. Simply said, the mean is an arithmetic average.

The sample median is also an important metric. The sample median's objective is to reflect the sample's core trend in a fashion unaffected by extreme results or outliers. A population's median is any figure less than or equal to half the population's values and greater than the values for the other half. It is not essential to know the value of extreme findings to compute a median because it is based on the middle data. Suppose we want to study to evaluate the time required to solve a problem. A median can still be derived if a small proportion of the participants fail to take the test in the allotted time.

There is a conceptual distinction between the mean and the median. The sample mean is the centroid of the data, whereas the median is always located halfway through the dataset. The difference between the median and the mean might be significant. However, the mean value for the no-phosphorous sample in the stem weight dataset is quite close to the median value.

1.4   Other Measures of Locations

Alternatives approaches provide values that compromise between the mean and the median. We do not utilize these other methods too often. However, the trimmed mean is a valuable metric. A trimmed mean is obtained by eliminating a percentage of the largest and smallest values in the dataset. For example, the 10% trimmed mean is computed by removing the greatest and smallest 10% of the values and finding the average of the remaining values.

As predicted, it is worth noting that the trimmed means for the individual samples are close to both the mean and median. Of course, the trimmed mean is less susceptible to outliers than the sample mean but not less than the median.

On the other hand, the trimmed mean strategy uses more data than the sample median. It is worth noting that the sample median is a variant of the trimmed mean, in which all of the sample data is removed except for the middle one or two observations.

1.5 Exercises

1. The following measurements were recorded for the time taken to complete a 200 m sprint in seconds (Table 1.3).
   Assume that the measurements are a simple random sample.

   (a) What is the sample size for the above sample?
   (b) Calculate the sample mean for these data.
   (c) Calculate the median of the sample.
   (d) Plot the data by way of a dot plot.
   (e) Compute the 20% trimmed mean for the above dataset.

2. The composite rods supplied by two separate companies were compared in a study. Twelve sample springs were manufactured using the rods given by each firm, and their flexibility was measured. The results are shown in Table 1.4.

   (a) Determine the data's sample mean and median for the two businesses.
   (b) On the same graph, plot the data for the two firms and outline any evident differences between the two businesses.

3. Research involving 16 adult males between 35 and 45 was conducted to see how a specific nutrition and exercise regimen affected blood cholesterol levels. Eight people were chosen randomly to be in the control group, while eight others were allocated to the therapy group over 12 months. Table 1.5 depicts the decrease in cholesterol observed in the 16 individuals over time.

   (a) Construct a dot plot for the data for both groups on one single graph.
   (b) For both groups, compute the mean, median, and 10% trimmed mean.
   (c) Explain why a variation in means leads to one conclusion regarding the effect of the regimen, whereas a difference in medians or trimmed means leads to an opposite conclusion.

4. Every day, the weather woman forecasts the day's maximum temperature. Loren uses her thermometer to record the day's highest temperature every day. As far as Loren can determine, the weather woman's forecast did not match the temperature

Table 1.3 Time taken to complete 200 m sprint

34	35	48	29	36
28	33	56	37	28
44	40	52	30	48

Table 1.4 Flexibility for the two companies

Company X	10.4	12.4	12.3	8.7	6.4	7.0	9.2	9.4	13.2	12.7	10.3	11.2
Company Y	11.4	13.5	12.4	12.7	9.6	11.5	11.5	13.5	12.2	12.6	8.5	11.0

Table 1.5 Cholesterol decrease for the two groups

Control group	9	− 3	4	12	1	23	5	− 9
Therapy group	− 7	31	17	12	12	13	− 3	4

values she obtained. What are the two variables that are being debated? Is there a positive, negative, or no association between the two?

1.6   Measures of Variability

In data analysis, sample variability is essential. Process and product variability are unavoidable in engineering and scientific systems, but controlling or reducing process variability can sometimes cause significant difficulties. Process engineers and managers increasingly realize that product quality and, as a result, earnings from produced goods are highly dependent on process variability. Material, environment, equipment, and other changes during the process are the root cause of variance for a stable process.

Even in minor data analysis situations, the effectiveness of a statistical approach may be determined by the level of variation among the sample's observations. Measures of sample location do not offer a complete picture of the nature of a dataset. For example, in Example 1.1, we cannot conclude that phosphorus improves seedling growth without accounting for sample variability.

1.6.1   Sample Standard Deviation and Sample Range

There are several measures of dispersion or variability, just as there are numerous measures of central tendency or location. The sample range $x_{max} - x_{min}$ maybe the most basic. The sample standard deviation is the most often used measure of spread. Let $x_1, x_2, \ldots, x_n$ signify sample values.

The sample variance, denoted by σ^2, is given by:

$$\sigma^2 = \frac{\sum_{n=1}^{i}(x_i - x)^2}{n - 1}.$$

The sample standard deviation, denoted by σ, is the positive square root of σ^2, that is,

$$\sigma = \sqrt{\sigma^2}.$$

Greater values of $(x_i - x)^2$ and large sample variance result from high variability in a data collection. The degrees of freedom associated with the variance estimate are called $n - 1$.

Example 1.3
A garden has a total of 25 plants. The following plants were chosen at random, and their heights in cm were recorded: 51, 38, 79, 46, and 57. Calculate the standard deviation of their heights.

$$n = 5, \text{ so, } n - 1 = 4$$

$$\text{Mean} (x) = \frac{(51 + 38 + 79 + 46 + 57)}{5} = 54.2$$

$$\text{Standard deviation} = \frac{\sum_{n=1}^{i}(x_i - x)^2}{n - 1} = 15.5$$

1.6.2 Units of Standard Deviation and Variance

The variance measures the average squared variation from the mean. Even though the concept uses a division by degrees of freedom of $n - 1$ rather than n, we use the phrase average squared deviation. Of course, if n is large, the denominator difference is insignificant. As a result, the sample variance has units equal to the square of the observed data units, while the sample standard deviation has linear units. Consider the data in Example 1.2 as an example. The weights of the stems are expressed in grams. Hence, the variance would be expressed in grams squared and the standard deviation in grams.

1.7 Exercises

1. A chemist wants to see if a pH meter has any "bias". The pH of neutral material (pH = 7.0) is measured on the meter to collect data. Eight samples are collected, and their pH values are shown in Table 1.6.

   (a) Calte the mean of the sample.
   (b) Calculate the variance and standard deviation of the sample.

2. Nadir conducted a study of his classmates' pet ownership, yielding the results (Table 1.7).

   (a) Calculate the mean number of pets owned.

Table 1.6 pH values obtained

6.54	7.32	7.15	6.99	6.56	7.98	8.03	6.12

Table 1.7 No. of pets owned

Number of pets	Frequency
0	5
1	13
2	7
3	0
4	4
5	7
6	2

Table 1.8 Salary of the colleagues

Salary (in dollars)	Frequency
3500	4
4000	6
4300	8
4700	3

    (b) What is the standard deviation?

    (c) What is the variance of this sample?

3. Table 1.8 shows the monthly salary of a group of colleagues in an advertising firm.

    (a) Determine the average salary.

    (b) Determine the standard deviation.

4. The standard deviation for the numbers 3, 8, 12, 17, and 25 is 7.563, accurate to three decimal places.

    (a) What happens if you multiply each of the five integers by five?

    (b) Explain how you arrived at your decision.

    (c) Determine the sample's variance.

1.8 Discrete and Continuous Data

Discrete data is a sort of numerical data that consists of numbers that have particular and definite values.

Many scientific fields employ statistical inference based on the analysis of observational data or controlled experiments. Depending on the application, the data collected might be discrete or continuous. A production engineer, for example, would be interested in running an experiment that results in optimal yield circumstances. Of course, the yield can be expressed as a percentage or as grams per pound on a scale.

Here are some instances of discrete data we may collect:

- The total number of consumers that purchased various things.
- Each department's total number of computers.
- The number of goods we buy each week at the grocery store.

Because discrete data is straightforward to summarize and analyze, it is frequently employed in elementary statistical analysis. Let us look at some of the other important features of discrete data:

- Discrete data consists of finite, numeric, countable, and non-negative integers with discrete variables (5, 10, 15, and so on).
- Simple statistical approaches such as bar charts, line charts, and pie charts can be used to depict and present discrete data.
- Discrete data can also be categorical, which means that it has a limited number of data values, such as a person's gender.
- Discrete data is distinctly spread in time and space. Discrete distributions make it easier to analyze datasets.

Continuous data is a type of numerical data that refers to the unspecified number of possible measurements between two realistic points.

The following are some instances of continuous data:

- Newborn infants' body weight.
- Daily wind speed.
- A freezer's temperature.

Continuous data, unlike discrete data, can be numeric or date and time distributed. This data type employs sophisticated statistical analysis techniques to account for the unlimited number of possible outcomes. The following are some of the most important aspects of continuous data:

- Continuous data evolves throughout time and can take on many forms at different times.
- Random variables, which may or may not be whole numbers, make up continuous data.
- Data analysis tools such as line graphs, skews, and others are used to measure continuous data.

- One of the most prevalent kinds of continuous data analysis is regression analysis.

Significant contrasts between discrete and continuous data in probability theory allow us to make statistical judgments. When the data is count data, statistical inference is frequently used. A physicist, for example, would be interested in counting the number of radioactive particles that pass through a counter every 100 μs. The number of oil tankers arriving each day at a particular port may be of interest to a logistic officer responsible for monitoring a port facility.

Several applications need statistical analysis of binary data. The sample percentage is frequently employed as a metric in analyses. For example, 40 patients suffering from a stomach condition were given medication. 10 out of 40 patients reported an improvement in their condition. Here, $10/40 = 0.25$ is the sample percentage for which the drug was successful, and $1 - 0.25 = 0.75$ is the sample proportion for which the drug was ineffective. In binary data, the basic numerical measurement is often indicated by either 0 or 1. A successful result is signified by a 1 in our clinical example, whereas a non-success is denoted by a 0. As a result, the sample percentage is a mean of ones and zeros.

1.9 Statistical Modeling, Scientific Inspection, and Graphical Diagnostics

Estimating the parameters of a speculated model is often the final result of statistical analysis. Data analysts, who frequently work with models, are aware of this. A statistical model cannot be completely deterministic; it must include some probabilistic elements. The analyst's assumptions are frequently built on the foundation of a model form. For example, in Example 1.2, the researcher could use sample data to make a model representing and predicting the variation of stem weight with changes in phosphorous level.

Statistical methods cannot entirely generate enough information or experimental data to define the population. However, datasets are frequently utilized to learn about population characteristics. Scientists and engineers are used to working with large amounts of data. The relevance of describing or summarizing the nature of data should be self-evident. A graphical depiction of a dataset may often provide information about the system from which the data was collected.

The importance of sampling and data visualization to improve statistical inference is examined in this section. We only present a few simple but helpful visual aids to studying statistical populations.

1.9.1 Scatter Plot

Dots indicate values for two separate numeric variables in a scatter plot (also known as a scatter chart or scatter graph). Scatter plots are used to see how variables relate to one another.

Example 1.4

- **Problem Statement**: Draw a scatter plot and show the relationship between the children's age, in years, and corresponding heights, in feet (Table 1.9).

- **Solution**

There are three simple steps to plot a scatter diagram.

Step I: Identify the x-axis and y-axis for the scatter plot.
Step II: Define the scale for each of the axes.
Step III: Plot the points based on their values (Fig. 1.4).

A line has been drawn through the points. We can see that a positive correlation exists between the two variables. As the variable on the x-axis increases, so does the variable on the y-axis. Thus, it can be deduced that height increases with the child's age.

The proposed model may take on a more sophisticated shape at times. Consider a clothing company that plans an experiment in which different percentages of wool are used to make fabric specimens. Take a look at the information in Table 1.10.

Example 1.5

- **Problem Statement**: In a company, six textile samples are made from each of the five wool percentages shown in Table 1.10. One purpose of this experiment may be to see which wool percentages are considerably distinct from the others.

Table 1.9 Age and height of the children

Age of the child	Height
3	2.3
4	2.7
5	3.1
6	3.6
7	3.8
8	4
9	4.3
10	4.5

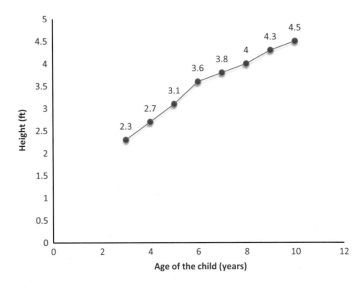

Fig. 1.4 Height versus age plot

	Wool percentage	Tensile strength
Table 1.10 Tensile strength for the varying wool content	20	13, 13, 14, 18, 17, 18
	25	19, 19, 19, 21, 21, 22
	30	22, 22, 19, 21, 20, 21
	35	24, 25, 26, 26, 25, 27
	40	10, 10, 11, 12, 11, 10

- **Solution**

The experiment's purpose often determines the type of model used to explain the data. The scatter plot shows that 25 and 30% wool textiles have similar strength; 20% wool gives an intermediate strength, while the strength for 35% wool is noticeably higher and that for 40% wool is significantly lower (Fig. 1.5).

Graphical representations may often highlight information that helps the data analyst understand the conclusions of the statistical inference. They can sometimes tell the analyst something that the formal analysis did not.

Almost every formal analysis needs assumptions derived from the data model. Violations of assumptions that may otherwise go unnoticed can be highlighted via graphics. Thus, graphics are utilized regularly throughout the text to enhance formal data analysis.

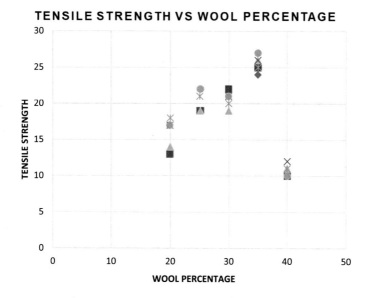

Fig. 1.5 Scatter plot of tensile strength versus wool percentages

1.9.2 Stem-and-Leaf Plot

When statistical data is shown in a stem-and-leaf plot, which combines tabular and visual displays, it may be beneficial to examine the distribution dynamics.

Stem-and-leaf diagrams are excellent for illustrating the relative density and type of data, providing a fast summary of the distribution to the reader. They keep (most) of the raw numerical data and may also be used to detect the mode and identify outliers. However, stem-and-leaf diagrams are only effective for modestly big datasets (about 15–150 data points). A stem-and-leaf display can be of limited utility with very small datasets, as a sufficient number of data points is necessary to establish conclusive distribution features.

Characteristics of an ideal stem-and-leaf plot:

1. A stem-and-leaf plot depicts the number's early digits (thousands, hundreds, or tens) as the stem and the last digit (ones) as the leaf.
2. Decimal points are rounded up to the next whole number. Test results, speeds, heights, and weights, for example.
3. When flipped on its side, it resembles a bar graph.
4. It depicts the data distribution, including the highest, lowest, most common, and outliers.

Example 1.6

- **Problem Statement**: A group of students in a class was asked to do a long jump and got the following results. Make a stem-and-leaf plot for the data provided (Table 1.11).

- **Solution**

See Table 1.12.

Note:

- Stem "3", Leaf "4" means 3.4.
- In this case, each leaf is a decimal.
- It is ok to repeat a leaf value.
- 5.0 has a leaf of 0.

Example 1.7

- **Problem Statement**: The stem-and-leaf plot of people who attended an election campaign from various locations is shown in Table 1.13.

  (i) What are the stem 4 data values?
  (ii) How many values are less than 19?

  Note: Stem 1 and leaf 3 represent 13.

- **Solution**

  (i) Stem 4 has two leaves that are 2 and 3. Hence, the data values are 42 and 43.

Table 1.11 Long jump results

2.2	2.5	2.4	3.4	2.9
3.5	2.3	2.0	1.9	3.6
5.0	4.9	1.9	2.4	4.3
2.3	2.7	3.4	3.8	3.7

Table 1.12 Stem-and-leaf plot for the long jump

Stem	Leaf	Frequency
1	9 9	2
2	0 2 3 3 4 4 5 7 9	9
3	4 4 5 6 7 8	6
4	3 9	2
5	0	1

Table 1.13 Stem-and-leaf plot for the campaign

Stem	Leaf
1	3 3 5
2	2 9
3	0
4	2 3

Table 1.14 Stem-and-leaf plot

Stem	Leaf
1	2 4
2	1 5 8
3	2 4 6
5	0 3 4 4
6	2 5 7
8	3 8 9
9	1

(ii) Looking for the corresponding leaf value of stem 1, we have 3, 3, and 5 in the leaf column. Thus, we find that the value less than 19 is 13, 13, and 15.

Example 1.8

- **Problem Statement**: Decipher the stem-and-leaf plot shown in Table 1.14 and use it to solve the problem.

  (i) Find out the mode of the plot.
  (ii) Calculate the mean of the plot.
  (iii) Determine the range.

- **Solution**

  (i) Mode is the number that appears most often in the data. Leaf 4 occurs twice on the plot against stem 5. Hence, mode = 54.
  (ii) The sum of all data values is $12 + 14 + 21 + 25 + 28 + 32 + 34 + 36 + 50 + 53 + 54 + 54 + 62 + 65 + 67 + 83 + 88 + 89 + 91 = 958$.

  To find the mean, we have to divide the sum by the total number of values.

  $$\text{Mean} = \frac{958}{19} = 50.42$$

  (iii) Range = Highest value − Lowest value = $91 - 12 = 79$.

A stem-and-leaf plot is a valuable tool for summarizing data. Another method is to utilize the frequency distribution, which involves counting the leaves on each stem and noting that each stem forms a class interval to divide the data into separate classes or intervals. Stem 1 in Table 1.6 denotes the interval 1.0–1.9, containing two observations. Similarly, stem 2 denotes the interval 2.0–2.9, which contains nine observations; stem 3 with six leaves denotes the interval 3.0–3.9.

1.9.3 Histogram

A histogram is a graph representing frequency distribution with continuous classes that have been categorized. It is a collection of rectangles with bases or intervals between class borders. The area of each bar is proportional to the frequency of the class. Histograms can reveal if data are not normally distributed.

The most common type of histogram is generated by dividing the data range into equal-sized bins (called classes). The horizontal axis represents the frequency (no. of samples belonging to each class), and the vertical axis represents the response variable.

The histogram graphically shows the following:

- Center of the data.
- Spad of the data.
- Skewness of the data.
- Presence of outliers.
- Presence of multiple modes in the data.

A distribution is considered symmetric if it can be folded along a vertical axis so that the two sides coincide. A skewed distribution lacks symmetry with respect to a vertical axis. The distribution seen in Fig. 1.6a is positively skewed since it has a long right tail and a considerably shorter left tail. Figure 1.6b shows that the distribution is symmetric, but Fig. 1.6c shows that it is negatively skewed.

Example 1.9

- **Problem Statement**: Mr. George is researching the height of the students studying at a university. He draws a histogram to analyze the height distribution of his students. Table 1.15 represents the sample data. The histogram is shown in Fig. 1.7.

- **Solution**

As shown in Fig. 1.7, a histogram with seven bins and seven frequencies is constructed. There is a height range on the x-axis. The number of students in each class is shown on the y-axis. Observing the third class, we can conclude that four students are 142–144 cm tall.

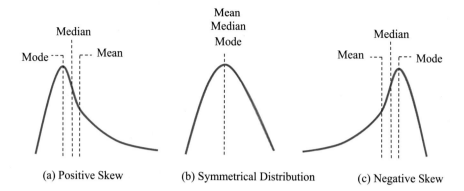

(a) Positive Skew   (b) Symmetrical Distribution   (c) Negative Skew

Fig. 1.6 Patterns of distribution

Table 1.15 Height of students

Serial No.	Height (cm)
1	141
2	143
3	145
4	145
5	147
6	152
7	143
8	144
9	149
10	141
11	138
12	143
13	145
14	148
15	145

Example 1.10

- **Problem Statement**: Table 1.16 of weight distribution among school students in a small town. Construct a histogram for the given data.

- **Solution**

See Fig. 1.8.

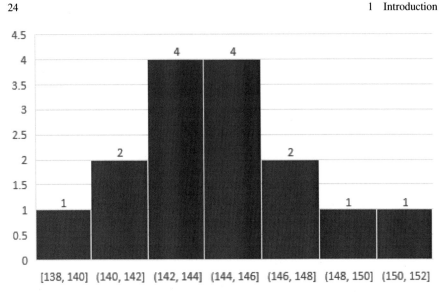

Fig. 1.7 Histogram representing the height distribution

Table 1.16 Weight distribution in the town	45	43	46	48
	55	59	56	54
	56	57	68	66
	75	77	74	73
	78	85	87	89
	95	95	93	98

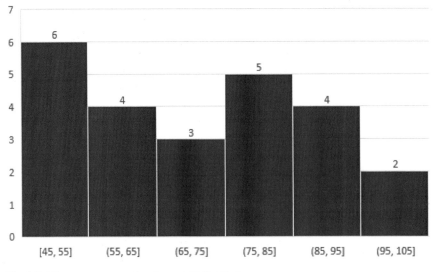

Fig. 1.8 Histogram representing the weight distribution

Table 1.17 Speed of the vehicles

20	22	24	26	29	31
34	33	35	43	44	48
50	52	53	55	62	64
71	73	74	80	86	82

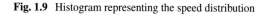

Fig. 1.9 Histogram representing the speed distribution

Example 1.11

- **Problem Statement**: Undergraduate students of transportation engineering at the Military Institute of Science and Technology observed the vehicle speed from a fixed point of Mohakhali Flyover. Draw a histogram for the data provided (Table 1.17).

- **Solution**

See Fig. 1.9.

1.9.4 Box-and-Whisker Plot or Box Plot

A box-and-whisker plot, often known as a box plot, shows a five-number summary of the sample set. The lowest value, first quartile, median, third quartile, and maximum value make up the five-number summary.

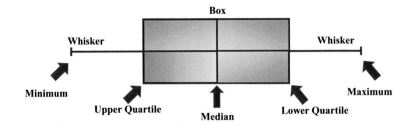

Fig. 1.10 Box-and-whisker plot

Fig. 1.11 Percentiles in a whisker-and-box plot

A box plot is created by drawing a box from the first to third quartiles. At the median, a vertical line runs through the box. The whiskers go horizontally from the lowest to the highest value, intersecting the quartiles. It effectively exposes a dataset's central tendency and variability, its distribution (especially symmetry or skewness), and the existence of outliers. It is also an effective graphical tool for contrasting samples taken from two or more groups.

So, there are a few things we should be aware of when working with box plots (Figs. 1.10 and 1.11):

(i) The smallest value in a dataset is called the lower extreme
(ii) The highest value in a dataset is called the upper extreme
(iii) The middle number in the collection is the median value
(iv) Lowest quartile—the lower 25% of the data is below that threshold
(v) Upper quartile—the upper 25% of the data is above that threshold
(vi) The lines that extend from the boxes are known as "whiskers". They are used to show how much variation exists between the upper and lower quartiles

Example 1.12

- **Problem Statement**: A sample of ten cartons of apples has the following weights (in grams):

38	29	28	29	30	35	34	35	37	25

Make a box plot of the data.

- **Solution**

Step 1: Sort the data in ascending order.

Hence, it becomes:

$$25, \ 28, \ 29, \ 29, \ 30, \ 34, \ 35, \ 35, \ 37, \ 38$$

Step 2: Calculate the median.

The median is the average of the two numbers in the middle.

$$25, \ 28, \ 29, \ 29, \ \mathbf{30}, \ \mathbf{34}, \ 35, \ 35, \ 37, \ 38$$

$$\frac{30 + 34}{2} = 32$$

Step 3: Determine the quartiles.

The median of the data points to the left of the median is the first quartile.

$$25, \ 28, \ \mathbf{29}, \ 29, \ 30$$

$$Q1 = 29$$

The median of the data points to the right of the median is the third quartile.

$$34, \ 35, \ \mathbf{35}, \ 37, \ 38$$

$$Q3 = 35$$

Step 4: Find the minimum and maximum values in the five-number summary.

The smallest data point is 25, which is the minimum.

The greatest data point is 38, which is the maximum.

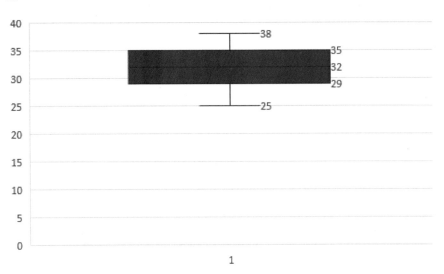

Fig. 1.12 Box-and-whisker plot for the given data

Thus, 25, 29, 32, 35, and 38 make up the five-number summary.

Step 5: Make the plot.

The five-number summary separates the data into parts containing 25% of the data in the sample.

About 75% of the cartons weighed more than 29 g (Fig. 1.12).

Example 1.13

- **Problem Statement**: A hi-tech company in Chittagong, Bangladesh, has two stores that sell personal computers. The company has recorded monthly sales of each store. In the past fiscal year, the following numbers of PCs were sold (Table 1.18).

  Draw the box-and-whisker plots for the two stores.

- **Solution**

First, we put the data points in ascending order.

20, 120, 160, 200, 210, 250, 290, 350, 380, 460, 510, 580.

Table 1.18 Sales of the two stores

Store 1	350	460	20	160	580	250	210	120	510	290	380	210
Store 2	520	260	380	180	500	630	420	210	70	440	140	80

Now, we need to find the median. However, this is an even set of data. The center lies between the sixth and the seventh values. So, we have to take an arithmetic average.

$$\frac{250 + 290}{2} = 270 \text{ is the median.}$$

Let us look at what happens in an even dataset with the lowest and higher quartiles. The following six figures are below the median: 20, 120, 160, 200, 210, and 250. The median of these six elements is the lower quartile; thus

$$\frac{160 + 200}{2} = 180 \text{ is the lower quartile.}$$

Six figures are higher than the median: 290, 350, 380, 460, 510, and 580. The median of these six data points is the upper quartile.

$$\frac{380 + 460}{2} = 420.$$

Finally, the sales totals for Store 1 are 20, 180, 270, 420, and 580.

We can obtain the five-number summary for Store 2 using the same calculations: 70, 160, 320, 470, 630 (Figs. 1.13 and 1.14).

Interpretation of results

The highest and lowest sales in Store 2 are more significant than the comparable sales in Store 1. Furthermore, the median sales value of Store 2 is more significant than Store 1. The interquartile range of Store 2 is also greater. As a result of these findings, we may conclude that Store 2 routinely sells more computers than Store 1.

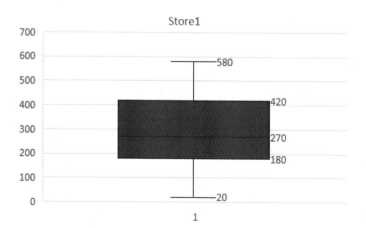

Fig. 1.13 Box plot for Store 1

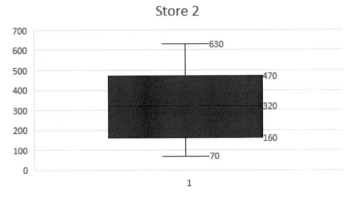

Fig. 1.14 Box plot for Store 2

1.9.5 Other Distinguishing Features of a Sample

Other characteristics of the distribution or sample can also describe its nature. While the median splits the data (or distribution) into two halves, there are additional metrics that can be highly useful in dividing segments or pieces of the distribution. The quartiles divide the data into four parts. By computing the percentiles of the distribution, the distribution may be split even further. The 95th percentile, for example, divides the top 5% of people from the lowest 95%. The 1st percentile is the line that divides the lowest 1% of the population from the rest of the population.

1.10 Exercises

1. The following is a list of the outage duration in seconds in a local power grid in Hanoi, Vietnam (Table 1.19).

   (a) Calculate the mean and median of the power-failure durations.
   (b) Calculate the standard deviation of the sample.

2. Table 1.20 shows the life expectancy of 24 identical mechanical devices, measured to the tenth of a year.

   (a) Create a stem-and-leaf plot for the sample.
   (b) Create a relative frequency distribution.
   (c) Determine the sample mean, range, and standard deviation.

Table 1.19 Duration in seconds

29	23	197	36	93	91	103	156	182	155	120	15	98	21
29	178	134	170	136	82	28	173	25	21	134	221	183	140

3. The following list contains middle school teacher salaries (in pounds per pupil) for 24 schools in Nottingham, England (Table 1.21).

   (a) Calculate the sample mean and standard deviation.
   (b) Create a stem-and-leaf representation of the data.
   (c) Create a histogram of the data's relative frequency.

4. An investigation is being undertaken on the impact of alcohol use on sleeping habits. The time it takes to fall asleep, in minutes, is the metric used (Table 1.22).

   (a) For each group, calculate the sample mean.
   (b) For each group, calculate the sample standard deviation.
   (c) On the same set of axes, construct dot plots for both the drinkers and non-drinkers.
   (d) Discuss how drinking affects the amount of time it takes to fall asleep.

5. The following are the lifespan in hours of forty 60-W, 120-V internally frosted luminous lamps. Construct a box plot for these data (Table 1.23).

Table 1.20 Life expectancy

2.9	1.3	1.5	3.2	1.4	1.9
1.3	3.0	5.0	3.2	3.4	2.4
1.2	4.3	2.0	2.9	2.0	3.9
1.5	1.0	2.4	2.9	3.2	4.9

Table 1.21 Salaries in pounds per pupil

3.87	1.08	2.45	3.33
2.34	2.43	0.90	2.43
3.56	4.34	1.34	3.90
1.09	2.33	1.00	2.56
3.03	3.29	3.54	4.05
4.03	1.22	0.98	1.34

Table 1.22 Sleeping time in minutes

Drinkers	17.3	22.6	21.8	28.6
	28.9	31.9	13.6	32.6
	36.8	23.9	14.9	12.8
Non-drinkers	55.2	67.3	34.9	44.3
	30.5	62.9	61.0	53.4
	71.2	45.0	60.6	59.3

Table 1.23 Lifespan of the bulbs

935	1187	987	935	811
1145	934	754	1036	1082
1176	964	830	1085	1145
1156	972	827	1289	992
987	953	973	1131	905
1246	855	755	1097	1272
786	900	885	972	1102
1232	1034	1067	1322	939

6. The histogram shows the heights of 21 students in a class, grouped into 5-inch-wide classes (Fig. 1.15).

   (a) How many students were greater than or equal to 55 inches tall but less than 70 inches tall?
   (b) Calculate the mean height of the sample.

7. Zorro is a swimmer, training for the Olympics. The number of 50-m laps she swam each day for 30 days is shown in Table 1.24.

   Prepare a stem-and-leaf plot. Make a brief comment on what it shows.

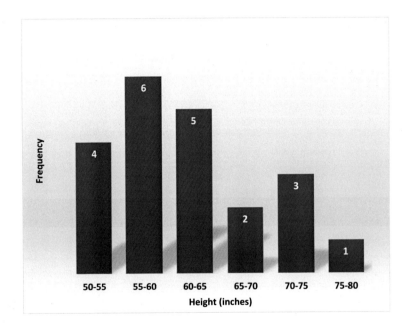

Fig. 1.15 Histogram representing height of students

Table 1.24 Number of laps

36	33	28	27	20	39	20	16	25	24
29	38	26	28	18	32	29	19	21	20
23	20	24	18	27	28	24	27	29	28

8. The following statistics are given for a sample of 40 measurements.
   25 is the maximum number.
   7 is the minimum.
   Quartiles are 9 and 17.
   $x = 453$ and $x^2 = 5803$, respectively.

   (a) Determine the mean and standard deviation of the data.
   (b) Given that the median is 11.2 create a box-and-whisker plot to display the data.
   (c) Assume that each number has been multiplied by 5. What are the new mean and standard deviation?

9. Draw a scatter plot for the given data that shows the number of games played and scores obtained in each instance (Table 1.25).

10. The population of different types of animals was noted in the Albuquerque Biological Park, New Mexico (Table 1.26).

    (a) Draw a scatter plot for the number of each animal listed in the zoo.
    (b) Is there any relationship that can be established between the two variables? If yes, how?

Table 1.25 No. of games played and points scored

No. of games	3	5	2	6	7	1	2	7	1	7
Score	80	90	75	80	90	50	65	85	40	100

Table 1.26 Animals in the park

Type of animal	No. of animals
Leopard	22
Lions	7
Monkeys	56
Elephants	12
Reindeer	31

Table 1.27 Temperature and humidity

Temperature (°F)	Humidity (%)
47	63
61	49
78	40
93	33
115	21
123	19

11. The Met Office in London has collected the following data about the temperature and humidity in the Greater London Built-up Area. Refer to Table 1.27 and indicate the method to find the humidity at a temperature of 60 °F.

    (a) Draw a scatter plot for the data.
    (b) Determine the line of best fit for the plot.
    (c) Determine the humidity at a temperature of 58 °F.

12. The following are the results of a math test for a class of 15 kids in John Marshalls High School, New York. The following are the outcomes:

82 84 89 93 72 83 93 90 76 69 77 89 92 100 70

    (a) Determine the median marks of the outcome.
    (b) Construct a box-and-whisker plot for the outcome.
    (c) What are your inferences from the plot you have drawn?

13. The following data is the weight of 20 students in a class. Construct a box plot with the following properties.

59 61 67 71 70 82 85 72 69 66
63 72 71 70 70 80 79 65 55 65

14. Atlanta's neighborhood ice cream store keeps track of how much ice cream they sell concerning the current day's midday temperature. Table 1.28 are their figures during the previous 12 days:

    (a) Construct a scatter plot for the sample.

Table 1.28 Temperature and ice cream sales

Temperature (°)	Ice cream sales
14.2	$215
16.4	$325
11.9	$185
15.2	$332
18.5	$406
22.1	$522
19.4	$412
25.1	$614
23.4	$544
18.1	$421
22.6	$445
17.2	$408

(b) Draw a line of best fit.

(c) From the graph, determine the ice cream scales at 16.9 °C.

15. The following table contains the heights of a group of students, measured in feet.

Height	No. of students	Height	No. of students
5	4	5.6	7
5.1	3	5.7	8
5.2	6	5.8	6
5.3	5	5.9	3
5.4	3	6.0	2
5.5	4	6.1	1

Find the mean, mode, and median of the student's heights. Also, determine the variance and standard deviation for the dataset.

16. The weight of fifteen bags of flour was measured in kilograms. The results are as follows: 2.00, 2.11, 2.05, 1.98, 1.96, 2.00, 2.03, 2.01, 1.94, 2.00, 1.93, 2.00, 1.99, 1.95, 2.00. What is the median weight of the bags of flour? Find the variance for the dataset.

17. The mean value of the following dataset is 7.04. Determine the value of q and the unknown frequencies.

X	2	4	6	8	10	12
Frequency	4	Q	3	$q+2$	5	3

18. Draw a box-and-whisker plot for the dataset shown below. Clearly label the median, the first and third quartiles, the maximum and the minimum values.

$$15, \ 45, \ 88, \ 90, \ 105, \ 93, \ 95, \ 86, \ 88, \ 100, \ 116$$

19. The heights of a group of people were measured in centimeters. The dataset below represents the results.

$$153, \ 182, \ 166, \ 162, \ 164, \ 180, \ 176, \ 183,$$
$$155, \ 178, \ 151, \ 186, \ 162, \ 175, \ 164, \ 177$$

$$172, \ 187, \ 179, \ 180, \ 196, \ 180, \ 186, \ 194,$$
$$191, \ 183, \ 175, \ 187, \ 166, \ 182, \ 198, \ 175$$

Plot a stem-and-leaf diagram to represent the above data.

20. There are 405 employees at a tech company. The following table shows the number of employees of different religions. Represent the information using a pie chart and also a bar chart.

Religion	Muslim	Christian	Hindu
Employees	250	35	120

References

Bowker, A. H., & Lieberman, G. J. (1972). *Engineering statistics* (2nd ed.). Prentice Hall.

Browne, R. H. (2001). Using the sample range as a basis for calculating sample size in power calculations. *The American Statistician, 55*(4), 293–298.

Devore, J. L. (2003). *Probability and statistics for engineering and the sciences* (6th ed.). Duxbury Press.

Gurland, J., & Tripathi, R. C. (1971). A simple approximation for unbiased estimation of the standard deviation. *The American Statistician, 25*(4), 30–32.

Hogg, R. V., & Ledolter, J. (1992). *Applied statistics for engineers and physical scientists* (2nd ed.). Prentice Hall.

Miller, I., Freund, J. E., & Johnson, R. A. (2000). *Miller and Freund's probability and statistics for engineers* (6th ed.). Prentice Hall.

Montgomery, D. C., & Runger, G. C. (2013). *Applied statistics and probability for engineers*. Wiley.

Ott, R. L., & Longnecker, M. T. (2000). *An introduction to statistical methods and data analysis* (5th ed.). Duxbury Press.

Walpole, R. E., & Myers, R. H. (1985). *Probability and statistics for engineers and scientists* (9th ed.). Macmillan.

Chapter 2
MATLAB

Abstract This chapter of "Statistics and Data Analysis for Engineers and Scientists" delves into the indispensable world of MATLAB, a powerful computational tool that serves as the linchpin of modern data analysis and scientific computation. This chapter offers readers a comprehensive initiation into MATLAB, encompassing fundamental concepts, essential commands, script and function files, arrays, and matrices. Readers will embark on a journey that equips them with the necessary skills for data manipulation, mathematical operations, and statistical analysis. Through a series of hands-on exercises and illustrative examples, this chapter instills proficiency in MATLAB, making it an indispensable asset for subsequent chapters' applications in statistical analysis and data interpretation. Beginning with an introduction to MATLAB, readers gain insight into the environment and its significance in engineering and scientific disciplines. They navigate through basic commands, empowering them to interact with the software efficiently. Script and function files are introduced, providing the framework for organizing and automating tasks, enhancing the reproducibility of analyses. Arrays and matrices, fundamental data structures in MATLAB, are explored in depth, enabling readers to handle complex datasets and perform advanced computations effortlessly. Practical examples illustrate the versatility of arrays and matrices in engineering and scientific applications. Moreover, readers will learn how to harness MATLAB's capabilities for mathematical operations, including algebraic functions such as solving linear equations and polynomial curve fitting. These skills are indispensable for addressing real-world problems that engineers and scientists encounter regularly. Incorporating data analysis and statistics, the chapter demonstrates MATLAB's prowess in managing and interpreting data. Readers will discover how to employ MATLAB for statistical tasks, facilitating data-driven decision-making and hypothesis testing. The chapter also unveils MATLAB's visualization prowess, with a focus on two-dimensional and three-dimensional plots. Through basic and specialized plotting techniques, readers will master the art of conveying complex data visually, a skill vital in engineering and scientific communication. Furthermore, this chapter introduces readers to numerical analysis, specifically the computation of roots of polynomials. This numerical capability empowers engineers and scientists to solve complex equations and model intricate phenomena effectively. By the end of this chapter, readers will have solidified their understanding of MATLAB, arming themselves with a powerful toolset to navigate the rich land-scape of data analysis, statistical computation, and scientific problem-solving that awaits them in subsequent chapters.

Keywords MATLAB · Introduction to MATLAB · Basic commands · Script files · Function files · Arrays and matrices · Mathematical operations · Relational operations · Algebraic functions · Linear equation solving · Polynomial curve fitting · Data analysis · Statistics · Two-dimensional plots · Three-dimensional plots

2.1 MATLAB

2.1.1 Introduction to MATLAB

MATLAB is a numerical calculation and visualization program. The abbreviation MATLAB stands for Matrix Laboratory. Hundreds of functions for technical analysis, graphical illustration, and animation are built into MATLAB. It also has a high-level programming language that allows for easy expansion.

The designed functions of MATLAB are used to handle data, do introductory algebra, solve differential equations (ODEs), and perform other scientific tasks. There are also various functions for plotting two and three-dimensional graphs. Users can even write their code using the MATLAB programming language. Besides that, MATLAB provides a platform for other coding languages such as C, Java, and Fortran. Another critical feature of MATLAB is the toolboxes that come with it. They cover various tasks, from financial and engineering statistics to image processing and even neural networks (Fig. 2.1).

The best feature of MATLAB is how simple it is to use. It also has a shallow learning curve, whereas the other methods have a relatively steep learning curve. Because MATLAB was explicitly designed for numerical computations, it is often much faster than C or Fortran at doing these calculations. MATLAB excels the most in numerical computations, particularly vectors and matrices, compared to other packages with similar objectives and scope.

MATLAB is powerful software. It takes some time to grasp its true potential. Unfortunately, such a powerful program can be intimidating to a newcomer. This section aims to help the readers overcome this fear and become productive in a short period.

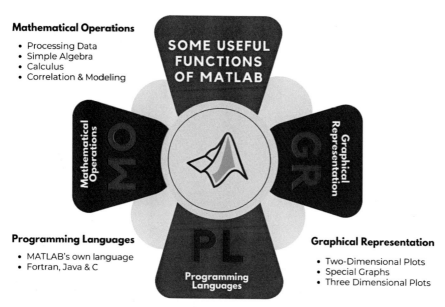

Mathematical Operations

- Processing Data
- Simple Algebra
- Calculus
- Correlation & Modeling

SOME USEFUL FUNCTIONS OF MATLAB

Mathematical Operations

Graphical Representation

Programming Languages

- MATLAB's own language
- Fortran, Java & C

Programming Languages

Graphical Representation

- Two-Dimensional Plots
- Special Graphs
- Three Dimensional Plots

Fig. 2.1 Basic functions of MATLAB

2.1.2 Some Basic Commands in MATLAB

MATLAB can be used as a calculator for simple arithmetic to even trigonometric operations. First, we will look at some basic operations using MATLAB before moving on to the more difficult sections.

Example 2.1

- **Problem Statement**

Suppose we want to add the following numbers: 10, 12, 23, 48, 54.

- **Solution**

In MATLAB, we must type the numbers in the command window with the addition (+) symbol between them. Next, we just press the enter key (Fig. 2.2).

Example 2.2

- **Problem Statement**

Now, suppose we want to subtract 2365 from 10,089. Again, we type the numbers in the command window with the subtraction symbol in between and hit enter.

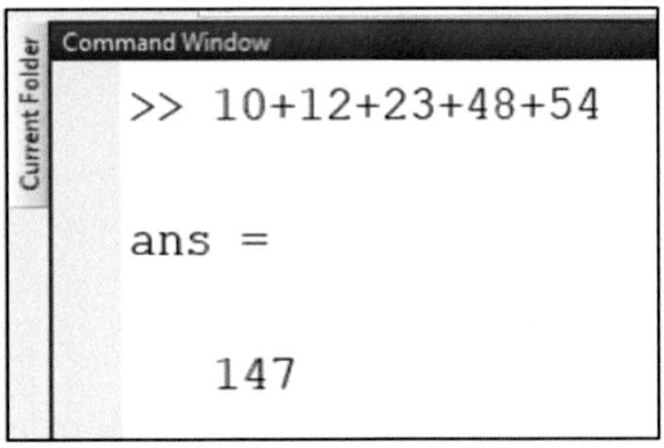

Fig. 2.2 Addition using
MATLAB

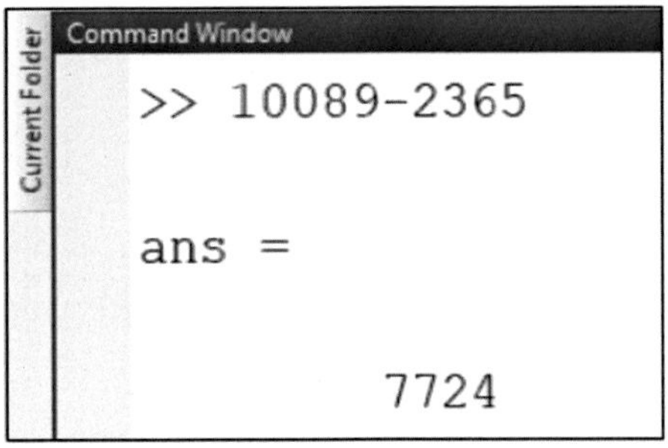

Command Window

>> 10+12+23+48+54

ans =

147

Fig. 2.3 Subtraction using
MATLAB

Command Window

>> 10089-2365

ans =

7724

- **Solution**

Again, we enter the numbers in the command window with the subtraction symbol
in between and hit enter (Fig. 2.3).

Example 2.3

- **Problem Statement**

If $x = 12$, $y = 8$, $z = 6$, find out the product of x and y, divided by z.

- **Solution**

First, we must assign the variables x, y, and z. Next, we find products x and y. Let us
call this variable a. Finally, we must divide a by z to obtain the result. We can write
the entire command in a single line without variables. A semicolon is used (end line)
to suppress the output of the instruction (Figs. 2.4 and 2.5).

Example 2.4

- **Problem Statement**

Let us take the number 1234 and represent it in various ways using the standard form.

Fig. 2.4 Assigning variables for multiplication and division

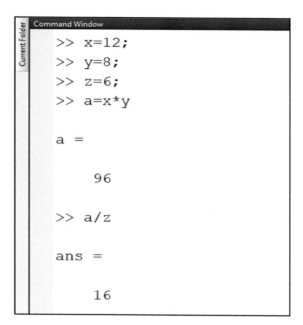

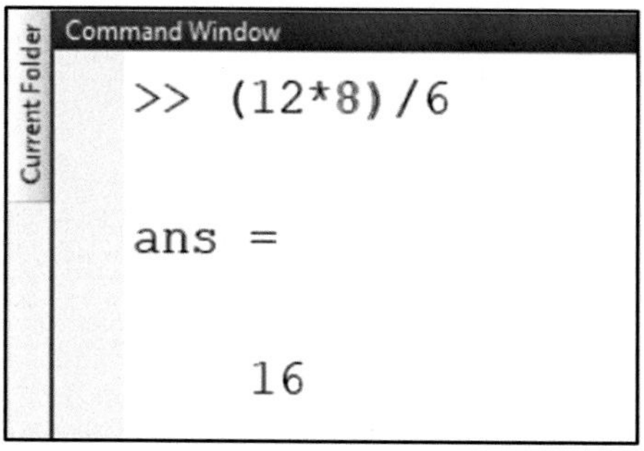

Fig. 2.5 Multiplication and division without variables

• **Solution**

This is done by writing an e to represent "multiplied by ten to the power", then adding a plus or minus to denote the symbol of the power, and finally writing the power itself. Next, we press enter (Fig. 2.6).

Example 2.5

• **Problem Statement**

Calculate the area of a circle, which has a radius of 3 units.

Current Folder
Command Window
>> 1234e-2
ans =
12.3400
Current Folder
Command Window
>> 1234e+2
ans =
123400
>> 1234e-12
ans =
1.2340e-09

Fig. 2.6 Writing numbers in standard form

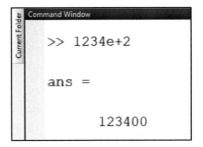

- **Solution**

See Fig. 2.7.

Current Folder
Command Window
>> r=3;
>> A=pi*r*r
A =
28.2743

Fig. 2.7 Area of the circle

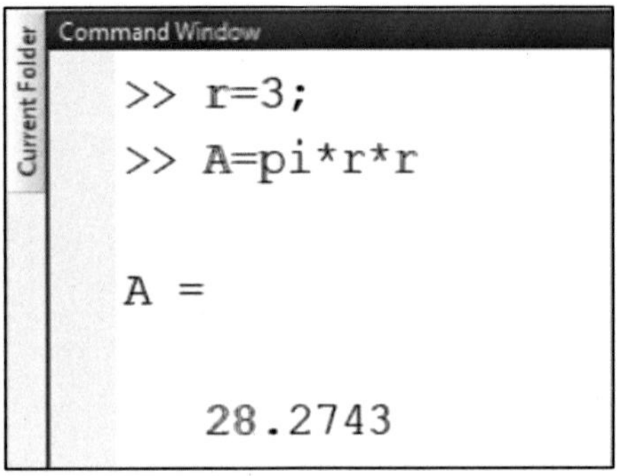

Example 2.6

- **Problem Statement**

Now, let's try the reverse and find the diameter of a circle, given an area of 36 units.

- **Solution**

See Fig. 2.8.

Example 2.7

- **Problem Statement**

Plot a circle with a radius of 1 unit.

- **Solution**

See Figs. 2.9 and 2.10.

Fig. 2.8 Diameter of circle

```
Command Window
>> A=36;
>> r=sqrt(A/pi)

r =

    3.3851

>> d=r/2

d =

    1.6926
```

Fig. 2.9 Command for plotting a circle

```
Command Window
>> theta=linspace(0,2*pi);
x=cos(theta);
y=sin(theta);
plot(x,y)
title('A Circle with 1 unit Radius')
xlabel('X'),ylabel('Y')
```

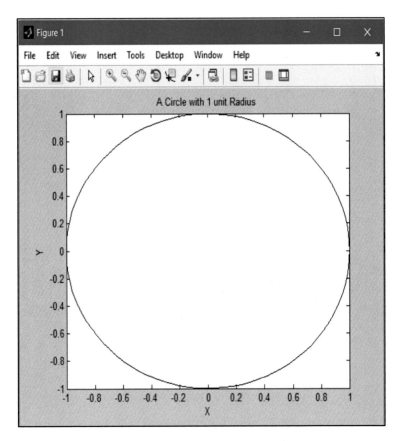

Fig. 2.10 Circle of unit radius

Example 2.8

- **Problem Statement**

Plot a tangent curve for a range of 0 to 2π.

- **Solution**

See Figs. 2.11 and 2.12.

Comment Lines

Comment lines are represented by "%". These are commands not meant to be executed by MATLAB. Comment lines describe the commands so that the primary writer and an editor can easily use them to make changes.

```
Command Window
>> x=linspace(0,2*pi);
plot(tan(x))
xlabel('x'),ylabel('tan(x)')
title('Tan(x) Curve')
grid on
```

Fig. 2.11 Command for plotting a tangent curve

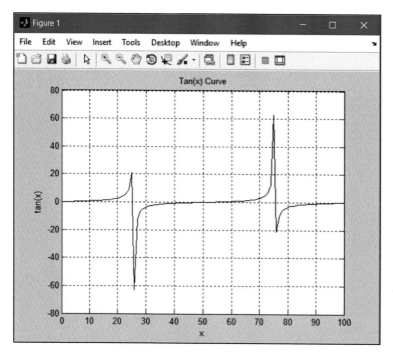

Fig. 2.12   Tangent curve

2.2   Script Files

A script file is a separate entity that stores command in text form, and all the commands in the script can be executed by referring to the script file name in the command window. Here, we have demonstrated how to make, store, and execute a script file in MATLAB.

Create a New File

File Menu → New → Script (Fig. 2.13).

Fig. 2.13 Creating a new script file

Save the File

File Menu → Save → Save As → Select Required Folder → Type the name of the file "Script1" → Save the file with the extension ".m" (Fig. 2.14).

The following command is used as an example. We can write any command we need. In the script file window, enter the following command (Fig. 2.15).

| Videos |
| Windows (C:) |
| RECOVERY (D:) |
| New Volume (E:) |
| New Volume (F:) |
| New Volume (G: |

File name: Script1

Save as type: MATLAB Code files (*.m)

e Folders   Save   Cancel

Fig. 2.14 Saving the script file with *.m extension

Fig. 2.15 Writing a command for a script file

```
Script1.m  ×  +
1 ─    x=linspace(0,4*pi);
2 ─    plot(x,sin(x))
3 ─    xlabel('x'),ylabel('sin(x)')
4 ─    title('Sine Curve')
5 ─    grid on|
```

Fig. 2.16 Execution of the script file

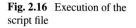

```
Command Window
fx >> Script1
```
Current Folder

Execute the Command

Save the changes made to the script file after entering the command. Enter the following in the command window, press Run from the menu above, or use the F5 function key from the keyboard. The script file will then be executed (Figs. 2.16 and 2.17).

2.3 Function Files

A function file contains an input argument and an output argument. Here is how we can create, save and execute a function file.

Create a New File

File Menu → New → Function (Fig. 2.18).

Save the File

File Menu → Save → Save As → Select Required Folder → Type the name of the file "CosCurve" → Save the file with the extension ".m" (Figs. 2.19 and 2.20).

Execute the Command

Save the changes made to the function file after entering the command. Enter the following in the command window, press Run from the menu above, or use the F5

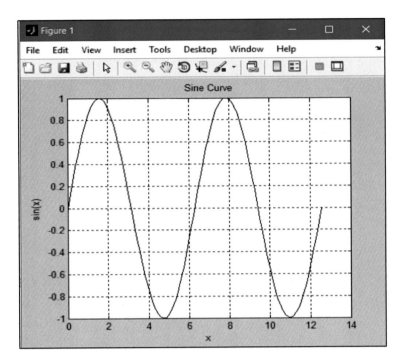

Fig. 2.17 Result of running the script file (sine curve plotted)

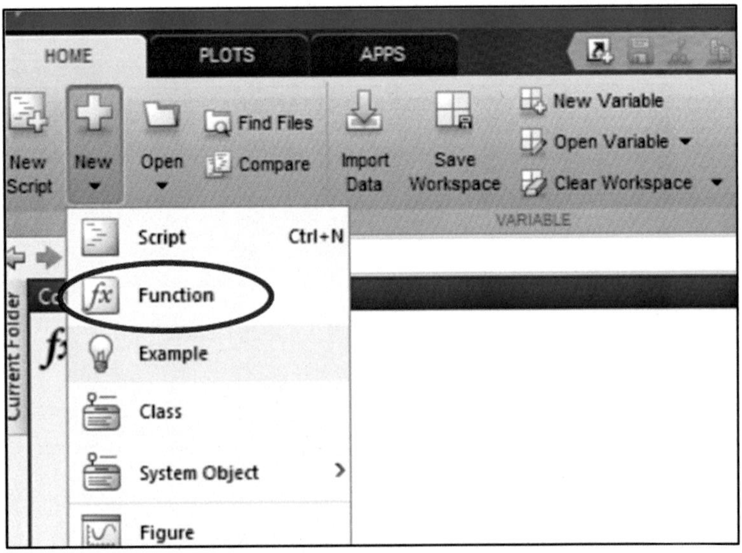

Fig. 2.18 Creating a new function file

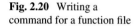

Fig. 2.19 Format of a function file

Fig. 2.20 Writing a command for a function file

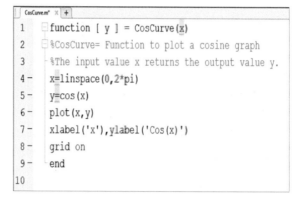

Fig. 2.21 Execution of the function file

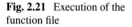

function key from the keyboard. The function file will then be executed (Figs. 2.21 and 2.22).

2.4   Arrays and Matrices

A sorted cluster of numbers may be classified into rows and columns. This representation is called an array. It may have one or more dimensions. A 2D array is referred to as a matrix. An array with "a" rows and "b" columns would be called an "$a \times b$"

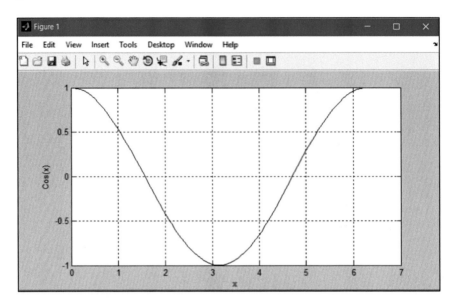

Fig. 2.22 Result of running the function file (cosine curve plotted)

matrix. MATLAB offers element-by-element operations for arrays and operations for entire matrices.

Suppose we want to enter a matrix M in MATLAB, where, M is as shown below.

$$M = \begin{matrix} 9 & 8 & 7 \\ 6 & 5 & 4 \\ 3 & 2 & 1 \end{matrix}$$

In order to enter a matrix in MATLAB, we have to type it in the following format. A single space separates different columns, whereas a semicolon separates different rows. After typing it in the command window, we have to press enter and MATLAB will represent the elements in a tabular form with rows and columns (Fig. 2.23).

$$M = [9\ 8\ 7;\ 6\ 5\ 4;\ 3\ 2\ 1].$$

Let's look at some basic operations using matrices in MATLAB (Figs. 2.24, 2.25, 2.26, 2.27 and 2.28).

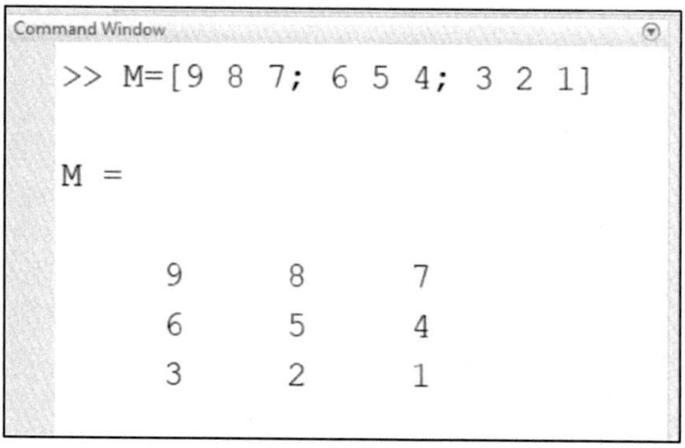
Command Window
>> M=[9 8 7; 6 5 4; 3 2 1]
M =
9 8 7
6 5 4
3 2 1

Fig. 2.23 Entering a matrix in MATLAB

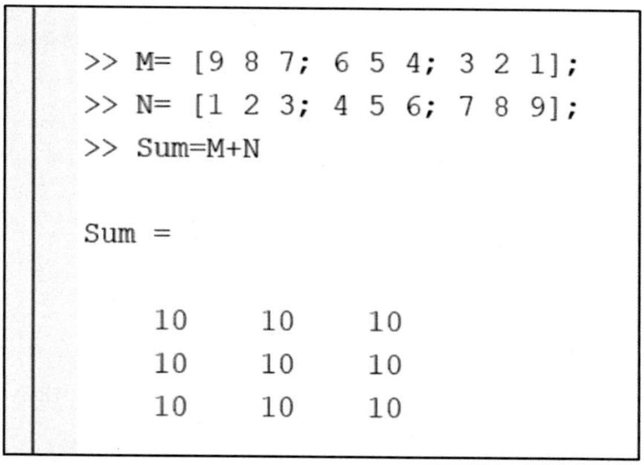
>> M= [9 8 7; 6 5 4; 3 2 1];
>> N= [1 2 3; 4 5 6; 7 8 9];
>> Sum=M+N
Sum =
10 10 10
10 10 10
10 10 10

Fig. 2.24 Adding two matrices directly in MATLAB

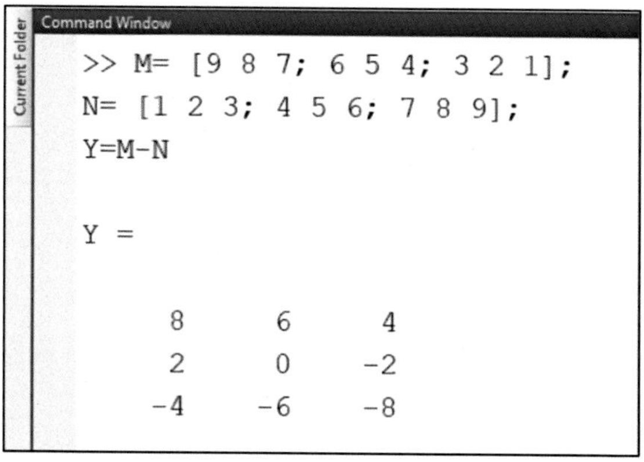
Current Folder
Command Window
>> M= [9 8 7; 6 5 4; 3 2 1];
N= [1 2 3; 4 5 6; 7 8 9];
Y=M-N
Y =
8 6 4
2 0 -2
-4 -6 -8

Fig. 2.25 Subtracting matrices directly in MATLAB

Fig. 2.26 Multiplying matrices directly in MATLAB

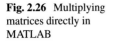

```
>> M= [9 8 7; 6 5 4; 3 2 1];
N= [1 2 3; 4 5 6; 7 8 9];
>> Product=M*N

Product =

        90     114     138
        54      69      84
        18      24      30
```

Fig. 2.27 Finding the transpose of a matrix

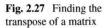

```
Command Window
>> A=[2 6 8; 4 1 0; 5 3 1]

A =

        2      6      8
        4      1      0
        5      3      1

>> transpose (A)

ans =

        2      4      5
        6      1      3
        8      0      1
```

2.5 Homework Problems

1. Compute the following using MATLAB as a calculator.

    (i) $\frac{6^2-6}{6}$

    (ii) $\frac{\sqrt{64}-4}{2}$

    (iii) e^2, ln e, log 1

2. Find the diameter of a circle with an area of 250 square inches.
3. Plot the following data on a set of axes and draw a best fit line. Label the axes and add a title to the plot.

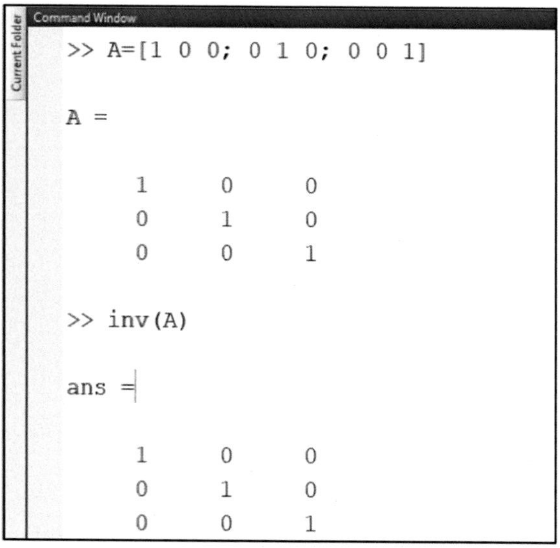

Fig. 2.28 Finding the
inverse of a matrix

```
Command Window

>> A=[1 0 0; 0 1 0; 0 0 1]

A =

        1       0       0
        0       1       0
        0       0       1

>> inv(A)

ans =

        1       0       0
        0       1       0
        0       0       1
```

X	2	4	6	8	10	12
Y	7	11	15	19	23	27

4. Create a function file for the equations and solve them. Execute the file in the MATLAB command window.

$$\text{If } x = 12, \quad y = 13, \quad z = 8$$

  (i) What is the value of $3x + 2z - y$?
  (ii) Find the value of $xyz - 4y/2z$.

5. Plot the following graphs on different axes.

$$\sin(4x) \quad \text{for } 0 < x < 2\pi$$

$$\cos\left(\frac{x}{2}\right) \quad \text{for } 0 < x < 2\pi$$

6. Perform the following matrix operations.

$$A = \begin{matrix} 9 & 8 & 7 \\ 6 & 5 & 4 \\ 3 & 2 & 1 \end{matrix} \qquad B = \begin{matrix} 9 & 8 & 7 \\ 6 & 5 & 4 \\ 3 & 2 & 1 \end{matrix} \qquad C = \begin{matrix} 9 & 8 & 7 \\ 6 & 5 & 4 \\ 3 & 2 & 1 \end{matrix}$$

(i) $AB + C$
(ii) $A^{\mathrm{T}}C - B$
(iii) $B^{-1}A + C.$

2.6 Mathematical Operations

2.6.1 Relational Operations

See Fig. 2.29.

2.6.2 Trigonometric Functions

See Fig. 2.30.

2.6.3 Other Trigonometric Functions

$$\sin, \ \ \sinh$$
$$a\sin, \ \ a\sinh$$
$$\cos, \ \ \cosh$$
$$a\cos, \ \ a\cosh$$
$$\tan, \ \ \tanh$$

2.7 Algebraic Functions

2.7.1 Solving a Linear Equation

Below is a system of linear equations.

Step-1 Every equation should be written with the unknown variables on the left side and the constants on the right side.

Fig. 2.29 Relational
operations in MATLAB

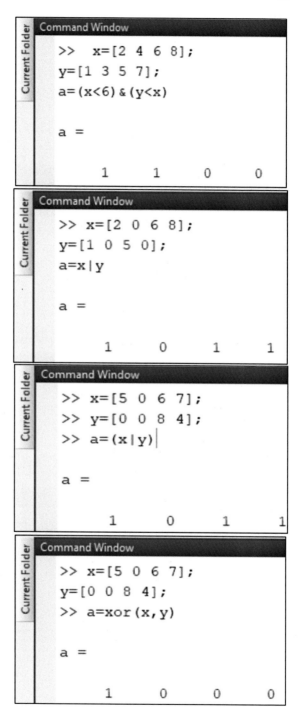

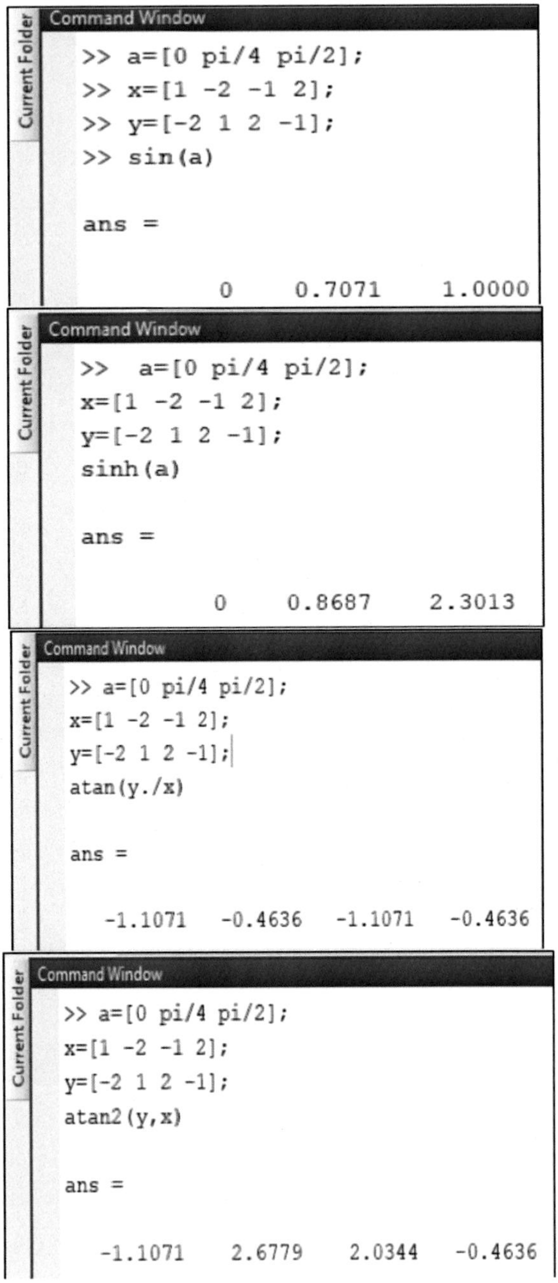
Current Folder
Command Window
>> a=[0 pi/4 pi/2];
>> x=[1 -2 -1 2];
>> y=[-2 1 2 -1];
>> sin(a)
ans =
0 0.7071 1.0000
Current Folder
Command Window
>> a=[0 pi/4 pi/2];
x=[1 -2 -1 2];
y=[-2 1 2 -1];
sinh(a)
ans =
0 0.8687 2.3013
Current Folder
Command Window
>> a=[0 pi/4 pi/2];
x=[1 -2 -1 2];
y=[-2 1 2 -1];
atan(y./x)
ans =
-1.1071 -0.4636 -1.1071 -0.4636
Current Folder
Command Window
>> a=[0 pi/4 pi/2];
x=[1 -2 -1 2];
y=[-2 1 2 -1];
atan2(y,x)
ans =
-1.1071 2.6779 2.0344 -0.4636

Fig. 2.30 Some examples using trigonometric functions

$$4x - 2y + 3z = 20$$
$$-5x + 6y + 8x = 12$$
$$4x + 3y - 2z = 4$$

Step-2 In matrix form, the equation should be written in the form $[A]x = b$, where x represents the unknown variables, A represents the coefficients of the unknown variables, and b represents the constants A, x and b are all vectors.

Here,

$$[x] = \begin{matrix} x \\ y \\ z \end{matrix}$$

$$[A] = \begin{matrix} 4 & -2 & 3 \\ -5 & 6 & 8 \\ 4 & 3 & -2 \end{matrix}$$

$$[b] = \begin{matrix} 20 \\ 12 \\ 4 \end{matrix}$$

Step-3 Type the command in MATLAB and press enter (Fig. 2.31).

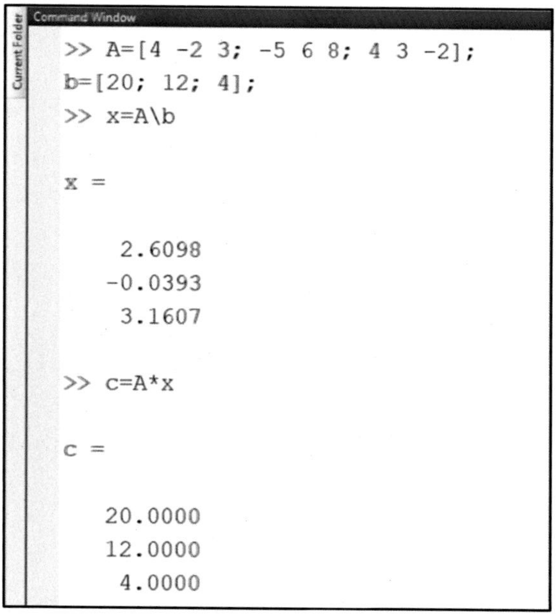

Fig. 2.31 Linear equation

```
Command Window

>> A=[4 -2 3; -5 6 8; 4 3 -2];
b=[20; 12; 4];
>> x=A\b

x =

    2.6098
   -0.0393
    3.1607

>> c=A*x

c =

   20.0000
   12.0000
    4.0000
```

2.7.2   Polynomial Curve Fitting

Enter the data and plot a scatter diagram (Figs. 2.32 and 2.33).

From the Tools option, select "Basic Fitting". In the window that appears, select the "linear" option (Figs. 2.34 and 2.35).

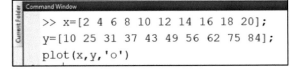

Fig. 2.32 Entering data for plot

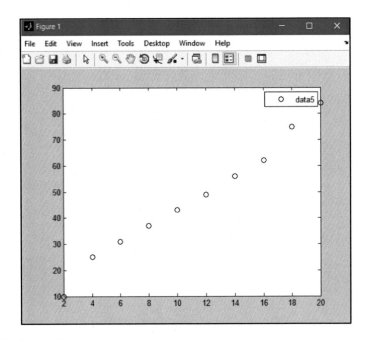

Fig. 2.33 Scatter plot

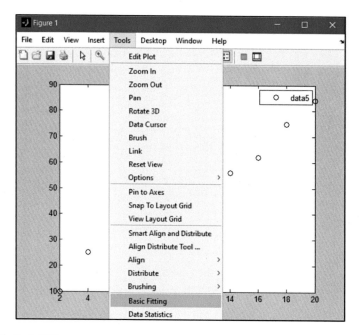

Fig. 2.34 Basic Fitting

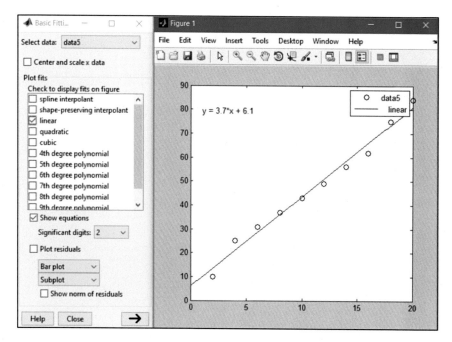

Fig. 2.35 Best fit line

```
>> m=[5 10 20 50 100];
d=[15 32 54 140 300];
g=9.81;
f=m/1000*g;
a=polyfit(d,f,1);
d_fit=0:10:300;
F_fit=polyval(a,d_fit);
plot(d,f,'o',d_fit,F_fit)
xlabel('Displacement\delta(mm)'),ylabel('Force(N)')
k=a(1);
text(100,0.32,['\leftarrow Spring Constant K=',num2str(k),'N/mm']);
```

Fig. 2.36 Command for plot

- **Example 5.3 Straight-line fit**

The following information was gathered during an experiment to determine a spring's spring constant.

$$F = k\delta$$
$$F = mg.$$

K can be found from this relationship as $k = mg/\delta$.

m (g)	5	10	20	50	100
Δ (mm)	15.5	33.07	53.39	140.24	301.03

The polynomial coefficients $a1$ and $a0$ must be found to fit a straight line through the data. Type in the following command in MATLAB and run it (Figs. 2.36 and 2.37).

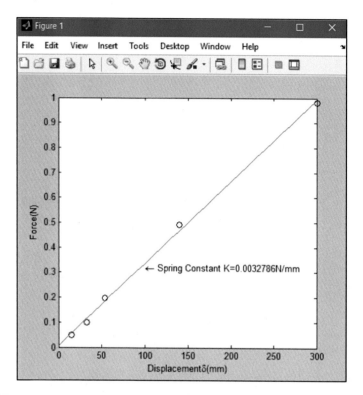

Fig. 2.37   Force versus displacement plot

2.8   Data Analysis and Statistics

We have already discussed basic statistical indices in Chap. 1. Let us try a few simple statistical operations using MATLAB, using the following dataset.

$$X = [200 \ 275 \ 320 \ 248 \ 330 \ 315 \ 287 \ 408 \ 297]$$

- **Mean**

See Fig. 2.38.

- **Median**

See Fig. 2.39.

- **Standard Deviation**

See Fig. 2.40.

- **Maximum Value**

Current Folder
Command Window
>> X=[200 275 320 248 330 315 287 408 297];
>> mean(X)
ans =
297.7778

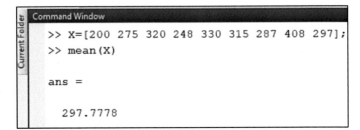

Fig. 2.38 Command for mean

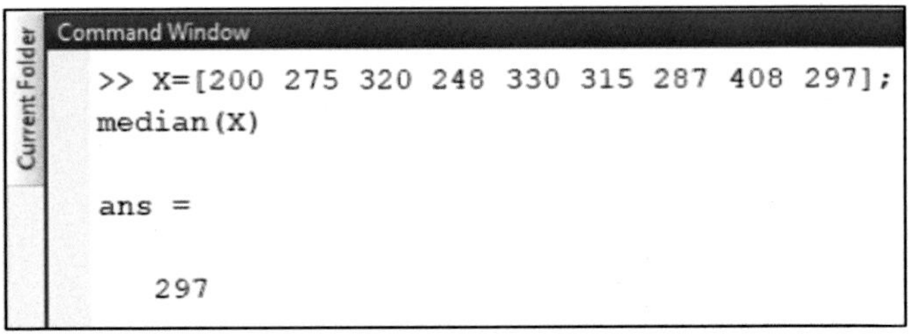

Current Folder
Command Window
>> X=[200 275 320 248 330 315 287 408 297];
median(X)
ans =
297

Fig. 2.39 Command for median

See Fig. 2.41.

- **Minimum Value**

See Fig. 2.42.

- **Sum**

See Fig. 2.43.

- **Percentiles**

See Fig. 2.44.

- **Box Plot**

See Fig. 2.45.

2.9 Two-Dimensional Plots

2.9.1 Basic 2D Plots

Plot (*x* values, *y* values, "style-option") is the most basic and useful command for making a 2D plot.

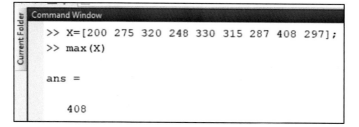

Fig. 2.40 Command standard deviation

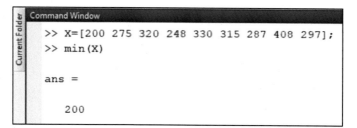

Fig. 2.41 Command for maximum value

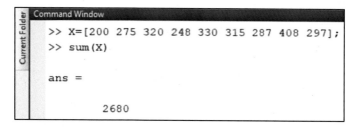

Fig. 2.42 Command for minimum value

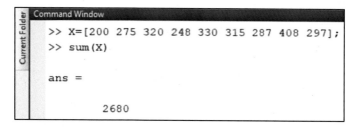

Fig. 2.43 Command for sum

```
Command Window
>> X=[200 275 320 248 330 315 287 408 297];
prctile(X,25)

ans =

  268.2500

>> prctile(X,75)

ans =

  322.5000
```

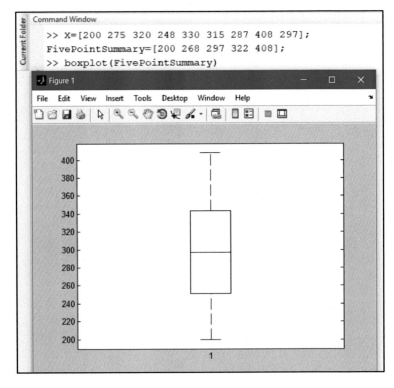

Fig. 2.44 Command for percentiles

Fig. 2.45 Box plot

Where x values and y values are vectors holding the x- and y-coordinates of graph points, style-option is an optional input defining the color, line style, and point marker style. Both the x value and y value vectors must be the same length. A single-vector parameter can also be used with the plot function. Thus, given two n-dimensional column vectors x and y,

Plot (x, y) y versus x; a solid line.
Plot $(x, y, \text{'—'})$ y versus x; dashed line.
Plot (x) plots x in relation to their row indices.

2.9.2 Modifying Plots with the Plot Editor

- For changing an existing plot, MATLAB provides an expanded interactive tool. To use this tool, go to the figure window and click the menu bar's left-leaning arrow. To modify any current plot object, pick and double-click it. When we double-click on an object, a property editor box appears, allowing us to pick and adjust the item's current properties.

- **Overlay plots**

We may make overlay charts in MATLAB using three separate commands: plot, hold, and line.

The hold command is another approach to creating overlay plots. When we use "hold on" during a session, the current plot in the graphics window is frozen. The plot command generates additional plots that are appended to the existing plot.

When the whole dataset to be plotted is not accessible simultaneously, the hold command is helpful for overlay plots (Figs. 2.46 and 2.47).

Fig. 2.46 Command for overlay plot

```
Command Window
fx >>   x=linspace(0,2*pi,100);
    y1=sin(x);
    plot(x,y1)
    hold on
    y2=x;plot(x,y2,'--')
    y3=x-(x.^3)/6+(x.^5)/120;
    plot(x,y3,'o')
    axis([0 5 -1 5])
    hold off
```

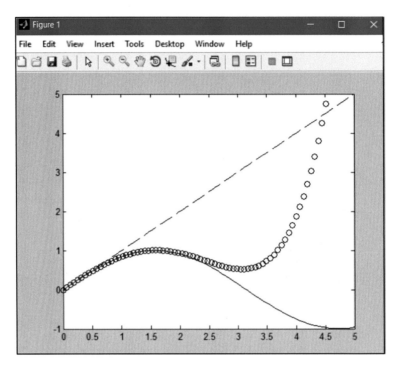

Fig. 2.47 Overlay plot

2.9.3 Specialized 2D Plots

For 2D charting, there are specific graphics operations. They can be used instead of the plot command we just covered. Other ex-plotter routines, such as ezplot, ezpolar, and ezcounter, are quite simple to use.

Other functions widely used for charting x–y data are listed below:

- **Pie Chart**

Suppose we want to draw a pie chart for cupcake sales in a bakery for the whole month. The sales of the four categories are as follows: Chocolate 432, Strawberry 235, Orange 184, Vanilla 308. Figure 2.48 shows the command for such a pie chart.

The figure shows the default pie chart obtained using the above command. Now, say we want to edit the pie chart manually. Select "Figure Palette" from the View option above the chart. After opening the Figure Palette, double-click on any segment of the pie chart to open the property editor for the figure (Figs. 2.49, 2.50 and 2.51).

- **Bar Chart**

See Figs. 2.52 and 2.53.

- **Horizontal Bar Chart**

```
Command Window
>> sales=[423;235;187;308];
pie(sales)
labels={'Chocolate','Strawberry','Orange','Vanilla'};
lgd=legend(labels);
title('Cupcake Sales','FontSize', 18);
```

Fig. 2.48 Command for pie chart

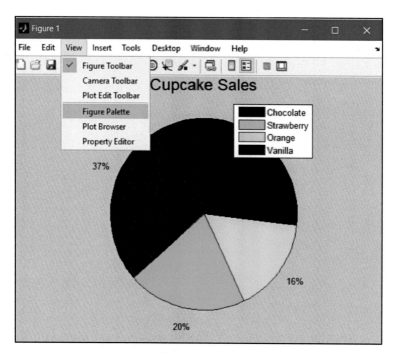

Fig. 2.49 Figure Palette

See Figs. 2.54 and 2.55.

- **Histogram**

Let us plot a histogram by generating random numbers between 1 and 50, and using the built-in function for the plot (Figs. 2.56 and 2.57).

- **2D Plot Using semilogx Function**

See Fig. 2.58.

- **2D Plot Using stem Function**

See Figs. 2.59 and 2.60.

Fig. 2.50   Editing the pie chart

Fig. 2.51   Final pie chart

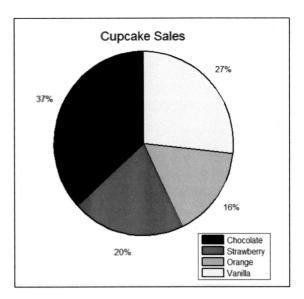

```
Command Window
>> y=[485 362 420 250 338];
bar(y)
set(gca,'XTickLabel',{'Math','Chemistry','Physics','Biology','Accounting'})
fx >>
```

Fig. 2.52 Command for bar chart

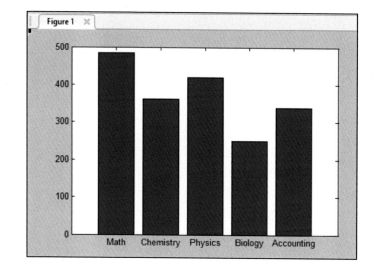

Fig. 2.53 Bar chart

Fig. 2.54 Command for
horizontal bar chart

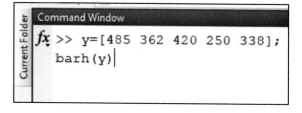

```
Command Window
fx >> y=[485 362 420 250 338];
barh(y)
```

2.10 Three-Dimensional Plots

MATLAB offers a broad scope for three-dimensional plotting. However, we will not
be exploring the details of these functions. We will illustrate a few examples to help
the readers understand mesh and surface plots.

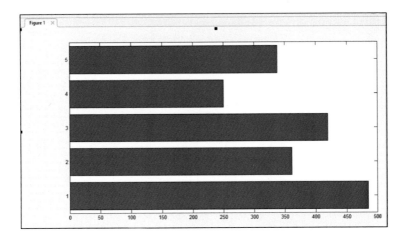

Fig. 2.55 Horizontal bar chart

Fig. 2.56 Generating
random numbers for
histogram

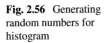

```
Command Window
>> x=randn(50,1);
>> hist(x)
>> colormap(summer)
```

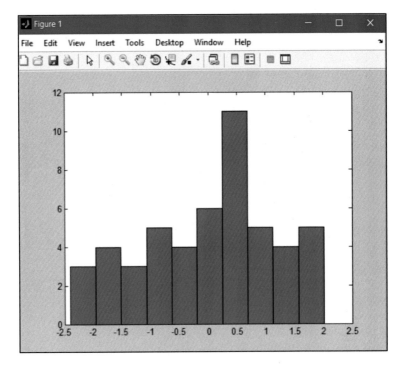

Fig. 2.57 Histogram

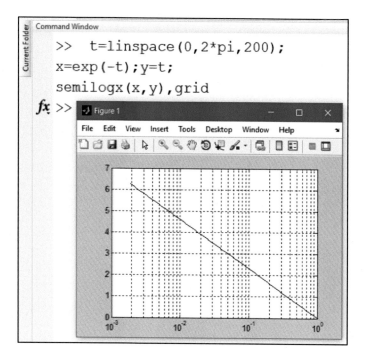

Fig. 2.58 Using the semilogx function

Fig. 2.59 Using the stem function

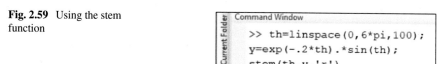

2.10.1 Mesh and Surface Plot Examples

See Figs. 2.61, 2.62, 2.63, 2.64, 2.65, 2.66, 2.67, 2.68, 2.69 and 2.70.

2.10.2 Modifying an Existing Plot

MATLAB produces handles for each item on the plot even if we do not explicitly create them when we construct the plot. We must first obtain the handle of any object before we alter it. We will need to know the parent–child connection between multiple graphical elements here. The following example shows how to acquire the handle of several objects and edit the plot using the handles.

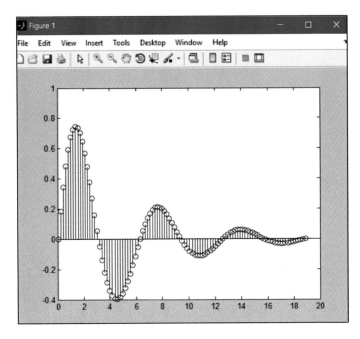

Fig. 2.60 Plot using the stem function

Fig. 2.61 Command for 3D plot

```
Command Window
>>   t=linspace(0,1,100);
x=t;y=t.^2;z=t.^3;
plot3(x,y,z),grid
xlabel('x(t)=t')
ylabel('y(t)=t2')
zlabel('z(t)=t3')
```

2.10.3 Loops

Control-flow expressions like for loops, while loops, and, of course, if-else branching have their syntax in MATLAB. It also has three commands to regulate the execution of scripts and functions: break, error, and return. These functions are described below.

- **For loops**

A for loop is a programming construct that repeats a statement or a set of statements a predetermined number of times.

Example

See Fig. 2.71.

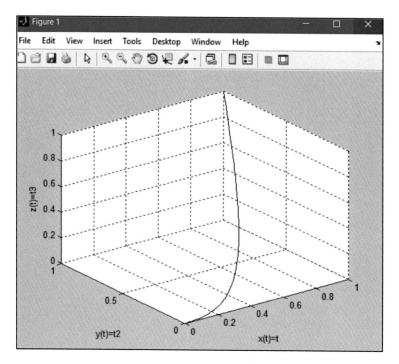

Fig. 2.62 Three-dimensional plot

Fig. 2.63 Command for
mesh plot

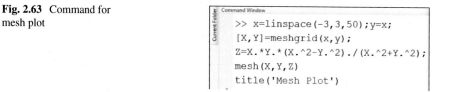

```
>> x=linspace(-3,3,50);y=x;
[X,Y]=meshgrid(x,y);
Z=X.*Y.*(X.^2-Y.^2)./(X.^2+Y.^2);
mesh(X,Y,Z)
title('Mesh Plot')
```

- **While loops**

A while loop executes a statement or a set of statements indefinitely until the condition
provided by while is no longer met (Fig. 2.72).

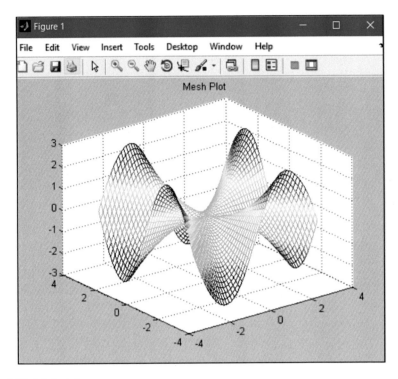

Fig. 2.64 Mesh plot

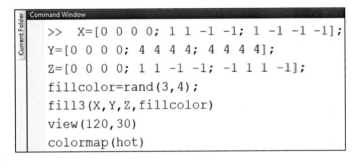

Fig. 2.65 Command filled polygons plot

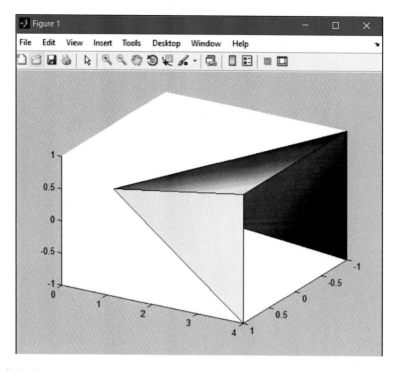

Fig. 2.66 Filled polygons with three vertices each

Fig. 2.67 Command for 3D stem plot

```
>> t=linspace(0,6*pi,200);
x=t;y=t.*sin(t);
z=exp(t/10)-1;
stem3(x,y,z,'filled')
```

2.11 Numerical Analysis

2.11.1 Roots of Polynomials

fzero may also be used to locate the zeros of a polynomial equation. However, fzero only finds the root that is closest to the initial assumption, not all of them. Use the built-in function roots to determine all roots of a polynomial problem.

$$x^5 - 3x^3 + x^2 - 9 = 0.$$

Here, $C = [1\ 0 - 3\ 1\ 0 - 9]$. As a result, the coefficient vector is always of length $n + 1$, where n is the polynomial order. In reality, MATLAB determines the

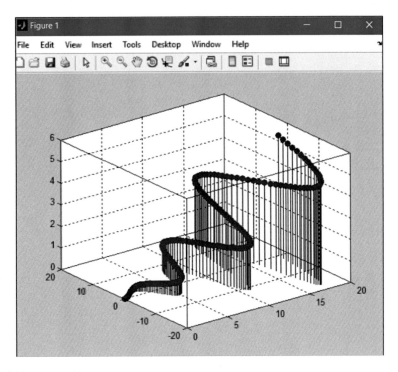

Fig. 2.68 3D stem plot

Fig. 2.69 Command for 3D
mesh plot

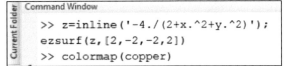

```
Command Window
>> z=inline('-4./(2+x.^2+y.^2)');
ezsurf(z,[2,-2,-2,2])
>> colormap(copper)
```

polynomial order based on the length of the vector. This is how you locate the roots
(Fig. 2.73).

2.12 Exercises

1. If $b = [0 \ \text{pi}/3 \ \text{pi}]$, $q = [3 - 3 - 3 \ 3]$, and $y = [3 \ 3 - 3 - 3]$, find the values of
   their

   (i)   sine
   (ii) hyperbolic sine
   (iii) tangent
   (iv) inverse tangent, and

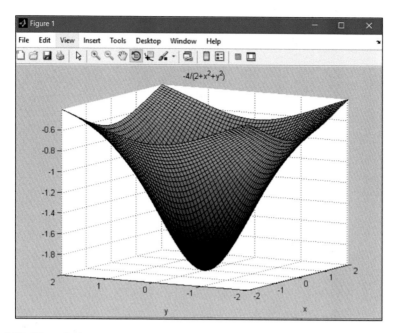

Fig. 2.70 3D mesh plot

   (v) cotangent.

2. If $d = [0\ 3pi/7\ 2pi]$, $q = [-1\ -1\ 1\ -1]$, and $y = [1\ -1\ -1\ 1\ 1]$, find the values of their

   (i) cosine
   (ii) inverse hyperbolic sine
   (iii) inverse cotangent
   (iv) hyperbolic tangent.

3. Solve these matrix equations using MATLAB:

$$6x - 3y + 2z = 9$$
$$5x - 7y + 8z = 11$$
$$7x - 4y - 9z = 22$$

4. Fitting linear and quadratic curves: You will be provided the following information.

$$X = 0,\ 0.17,\ 0.2,\ 0.34,\ 0.42,\ 0.53,\ 0.61,\ 0.72,\ 0.87,\ 0.95,\ 1.0$$
$$Y = 0,\ 0.71,\ 2.2,\ 3.46,\ 4.5,\ 6.34,\ 7.3,\ 9.16,\ 10.7,\ 12.8,\ 16$$

5. Calculate the values of x and y from the group of linear equations, as given.

Fig. 2.71 For loop

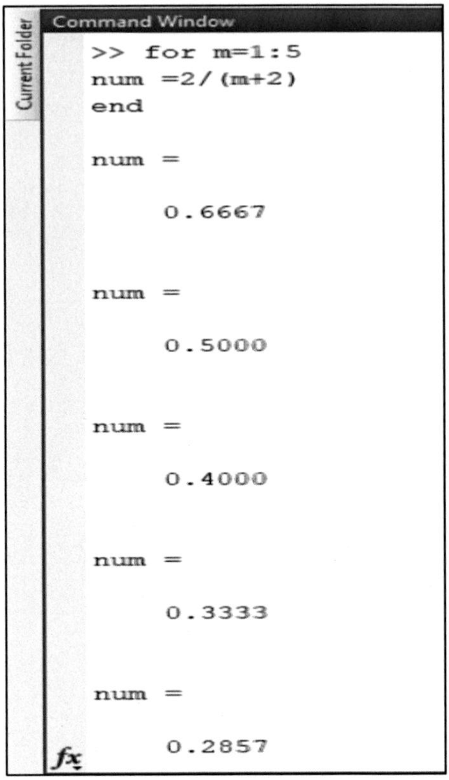

```
>> for m=1:5
num =2/(m+2)
end

num =

    0.6667


num =

    0.5000


num =

    0.4000


num =

    0.3333


num =

    0.2857
```

$$x + 2y + 3z = 1$$
$$3x + 5y + 4z = 1$$
$$2x + 3y + 3z = 2.$$

(i)   Write the equation in matrix form and use left division to solve for $x = [x\ y\ z]f$.

(ii) Find the answer again using the augmented matrix's rref function.

(iii) Can you discover the answer using the LU decomposition? Because $[LU]x = b$, let $[U]x = y$, resulting in $[L]y = b$. Now solve for y first, then for x.

Exercise 2.6

Fig. 2.72 While loop

```
Command Window
>> a=1;b=1;i=1;
while b<2000
a=[a;b];
i=i+1;
b=2^i;
end
a

a =

                    1
                    1
                    4
                    8
                   16
                   32
                   64
                  128
                  256
                  512
                 1024
```

Fig. 2.73 Roots of an equation

```
Command Window
>> c=[1 0 -3 1 0 -9];
>> roots(c)

ans =

   1.9316 + 0.0000i
   0.5898 + 1.1934i
   0.5898 - 1.1934i
  -1.5556 + 0.4574i
  -1.5556 - 0.4574i
```

1. Draw a 2D graph for the following function:

$$f(t) = t \cos t, \quad 0 \le t \le 5$$

2. Draw a two-dimensional plot for the function given: $x = e^{-m}, y = t$.
   The limit of m ranges from 0 to $360°$.

3. Plot the curve of the function

$$F = e^{-t/4} \cos t, \quad 0 \le t \le 360°$$

4. Draw a two-dimensional plot for the function given:

$$P = e^{-2x}, \quad y = 2x, \quad 0 \le x \le 720°$$

Chapter 3
Excel

Abstract This chapter of "Statistics and Data Analysis for Engineers and Scientists" plunges readers into the versatile realm of Microsoft Excel, a ubiquitous tool for data analysis, manipulation, and visualization. This chapter is a comprehensive guide to harnessing Excel's power for engineering and scientific applications. Beginning with an introduction to Excel, readers gain an understanding of its importance in modern data-driven disciplines. The fundamentals of Microsoft Excel are explored, including worksheets, workbooks, rows, columns, and cells. Vital data formatting techniques are introduced, ensuring data presentation clarity. Readers are guided through inputting functions and formulas directly into cells and using the formula bar, unleashing Excel's computational prowess. The chapter dives into essential mathematical operations, such as SUM, MIN, MAX, and SUMPRODUCT, bolstering data analysis capabilities. It also delves into trigonometric functions, covering units in radians and degrees and introducing SIN, COS, and TAN. Measures of location and variation, including mean, median, mode, variance, standard deviation, quartiles, percentiles, and box-and-whisker diagrams, are thoroughly explained. The chapter proceeds to explore correlation and regression modeling, elucidating Pearson's correlation coefficient, scatter diagrams, trend lines, linear and nonlinear models, and FORECAST. Graphical representation's significance in data analysis is highlighted, with practical guidance on creating pie charts, bar charts, histograms, and stem-and-leaf diagrams in Excel. Financial data analysis techniques are covered, encompassing currency formatting, simple and compound interest, number of payments (NPER and PDURATION), and depreciation using the straight-line method. The chapter concludes with exercises to reinforce newfound skills and encourage practical application. By the end of this chapter, readers will have acquired a robust understanding of Microsoft Excel's capabilities as a comprehensive tool for data manipulation, mathematical analysis, visualization, and financial modeling, providing them with a formidable skill set for engineering and scientific data-driven tasks.

Keywords Introduction to excel · Fundamentals of Microsoft Excel · Worksheets and workbooks · Rows, columns, cells · Data formatting · Inputting functions and formulas · Trigonometric functions · Mean, median, mode · Variance and standard deviation · Quartiles and percentiles · Box-and-Whisker diagram · Correlation and regression modeling · Pearson's correlation coefficient · Scatter diagrams and trendlines · Graphs and charts

3.1   Introduction to Excel

Of the three programs discussed in this book, Microsoft Excel is the most common and straightforward software for beginners. This section is dedicated to exploring the essential functions that Excel has to offer in terms of data analysis. It does not cover all possible operations, of course. However, it aims to provide a basic understanding of Excel so that the users can learn how to manoeuver their way through the software.

Excel workbooks are employed in various fields, including business, education, personal finance, and recreation. Excel is one of the business's most widely used analytical and reporting applications. Excel keeps track of such information, including accounting, sales, stock, project scheduling, and client activities. The application is alluring due to its capacity for data manipulation and feedback. Excel is wonderfully flexible with data storage and presentation (Fig. 3.1).

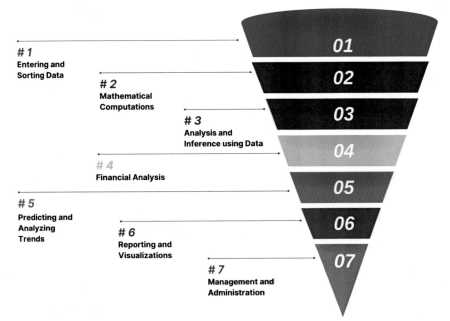

Fig. 3.1 Useful functions of Excel

3.2 The Fundamentals of Microsoft Excel

Microsoft Excel is the most widespread and user-friendly software for novices among the three programs outlined in this book. This section is dedicated to exploring the essential functions that Microsoft Excel has to offer in terms of data analysis. It does not cover all possible operations, of course. Nonetheless, it strives to impart a fundamental grasp of Excel so that users may learn how to use the program.

3.2.1 Worksheets and Workbooks

A Microsoft Excel file is known as a workbook. Just like a book with different pages, a workbook contains different pages called worksheets or spreadsheets. The purpose of different sheets is to divide our work into separate segments while keeping it all under the same file. Below, we can see a worksheet opened in a workbook (Fig. 3.2).

We can delete, rename, move, or copy the sheets by right-clicking on the sheet tab, and selecting the required option. To rename the tab directly, we can also double-click on the sheet tab (Fig. 3.3).

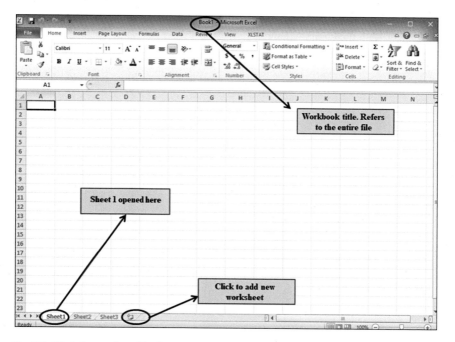

Fig. 3.2 Worksheet and workbook

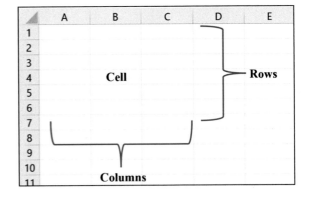

12	
13	Insert...
14	Delete
15	Rename
16	Move or Copy...
17	View Code
18	Protect Sheet...
19	
20	Tab Color ▶
21	Hide
22	Unhide...
23	Select All Sheets

◄ ◄ ► ►   Sheet1   Sheet2

Ready

Fig. 3.3 Move, copy, remove, or rename a sheet

3.2.2   Rows, Columns, Cells

A spreadsheet is a table with rows and columns. Each box on the table is called a cell, where the data is generally entered. We refer to a cell by its row and column number. For example, in the figure below, the marked cell would be called "B4" (Fig. 3.4).

Fig. 3.4 Rows, columns, and cells

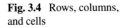

3.2.3 Data Formatting

As mentioned in the previous section, data is generally entered in the cells. Once we type in our data, we can format it using several options found right above the worksheet (Fig. 3.5).

In the figure below, if we write the titles "Number of Sales" and "Flavour of Cupcake", it does not fit within the cell's dimensions. For the first title, we see only part of it being displayed, whereas, for the second title, it has been extended onto the next cell. How can this be presented more properly? (Fig. 3.6).

The solution is simple. The alignment toolbar can be used to modify the alignment as needed. Here, we have used the "wrap text" option to display our text within the given cell width. The width can be adjusted by placing the cursor between the two column headings H and I and then dragging it to the left or right as needed. There are also options for adjusting text alignment—left, right, top, bottom, middle, or center (Figs. 3.7 and 3.8).

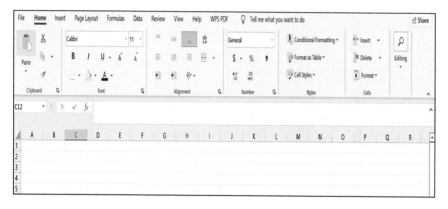

Fig. 3.5 Options for data formatting

Fig. 3.6 Non-formatted text

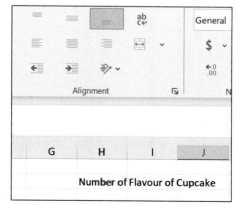

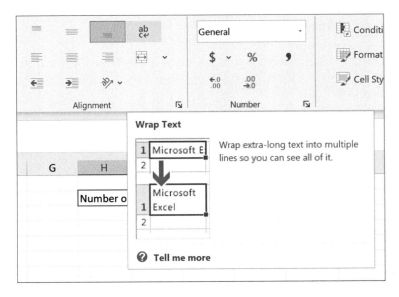

Fig. 3.7 Wrapping text

Fig. 3.8 Wrapped text

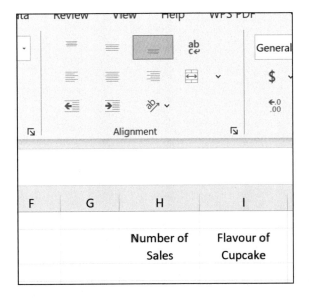

There are many options for formatting data (text or numbers) using the tools shown in Fig. 3.4. They can be explored individually as the user gets more accustomed to Excel. We will not be addressing every possibility here.

Fig. 3.9 Entering quiz scores

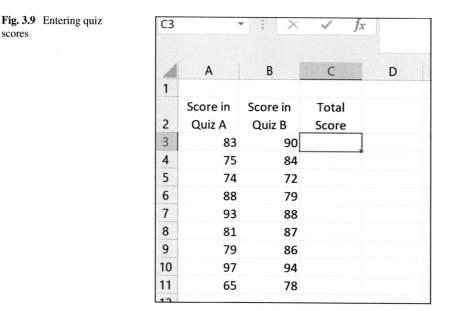

3.3 Inputting Functions and Formulas

There are several ways of applying formulas in Excel. We will be looking at two basic approaches in this section.

3.3.1 Directly in the Cells

Suppose a group of students took two tests, Quiz A and Quiz B. The sum of the scores obtained in the two tests determines the final result. In Excel, we can insert a formula directly into a cell. This same formula can be applied to all the cells in the column, or we can enter a different formula for each cell (Fig. 3.9).

To find the total score, we must first identify the cell numbers. For the first pair of scores, we have to enter our formula in cell C3. The two scores we need are in cells A3 and B3. Thus, our input will be as follows. After writing the formula, we can obtain the result by pressing enter (Figs. 3.10 and 3.11).

3.3.2 Using the Formula Bar

Similarly, the formula can be written using the formula bar above the worksheet. The result is obtained by pressing enter (Fig. 3.12).

Fig. 3.10 Entering formula in cell

	A	B	C	D
1				
2	Score in Quiz A	Score in Quiz B	Total Score	
3	83	90	=A3+B3	
4	75	84		
5	74	72		
6	88	79		
7	93	88		
8	81	87		
9	79	86		
10	97	94		
11	65	78		

Fig. 3.11 Result obtained from the current analysis

	A	B	C	D
1				
2	Score in Quiz A	Score in Quiz B	Total Score	
3	83	90	173	
4	75	84		
5	74	72		
6	88	79		
7	93	88		
8	81	87		
9	79	86		
10	97	94		
11	65	78		

The formula can be extended to the rest of our dataset. Place the cursor at the bottom corner of cell C3. A small black "+" symbol should appear. Click and drag down from that point, and Excel will apply the same formula format to the cells below. Note that we mentioned the same formula format rather than the formula itself. It implies that Excel detects a pattern in the formula, i.e., summation of the cells in columns A and B of the same row. Thus, for cell C4, the auto-generated formula will be "= A4 + B4" (Fig. 3.13).

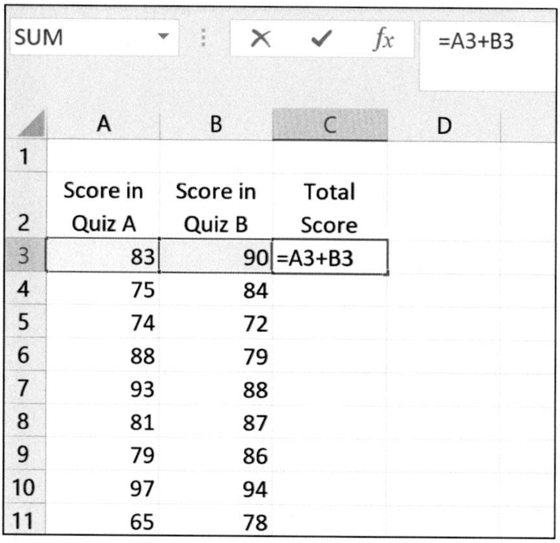

Fig. 3.12 Using the formula bar

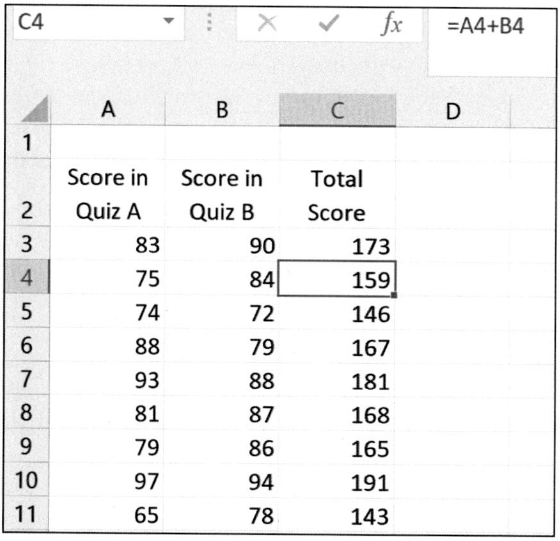

	A	B	C	D
1				
2	Score in Quiz A	Score in Quiz B	Total Score	
3	83	90	=A3+B3	
4	75	84		
5	74	72		
6	88	79		
7	93	88		
8	81	87		
9	79	86		
10	97	94		
11	65	78		

SUM ▾ ⋮ ✕ ✓ *fx* =A3+B3

Fig. 3.13 Applying the formula pattern to the remaining cells

C4 ▾ ⋮ ✕ ✓ *fx* =A4+B4

	A	B	C	D
1				
2	Score in Quiz A	Score in Quiz B	Total Score	
3	83	90	173	
4	75	84	159	
5	74	72	146	
6	88	79	167	
7	93	88	181	
8	81	87	168	
9	79	86	165	
10	97	94	191	
11	65	78	143	

3.4 Some Important Mathematical Operations

Excel has numerous mathematical operations that can be applied to small or large datasets. These operations enable easy and efficient computation, saving us time. This section introduces the readers to some of the basic mathematical operations that Excel offers.

C	D	E		C	D	E
	Dataset				**Dataset**	
	14				14	
	12				12	
	15				15	
	16				16	
	11				11	
	19				19	
	13				13	
	12				12	
	16				16	
	15				15	
SUM	=SUM(D2:D11					
	SUM(**number1**, [number2], ...)			**SUM**	143	

Fig. 3.14 Applying the SUM function

First, we have to enter our values into the worksheet. Next, type a "=" symbol followed by "OPERATION" and start a bracket. Inside the bracket, we have to add our data. Place the cursor on the cell containing the first value of the dataset. Press and hold. Drag it down to the last value. Close the bracket, press enter, or click on an empty cell to get the result. The word "OPERATION" will be replaced with the name of the operation you are using. Let us try out the following operations.

3.4.1 SUM

See Fig. 3.14.

3.4.2 MIN and MAX

See Figs. 3.15 and 3.16.

3.4.3 SUMPRODUCT

Figure 3.16 represents the goods sold at a bakery on a particular day and the unit price of each item. How can we calculate the total amount of revenue from the sale of these goods? To find the revenue from a particular item, we would have to multiply

F	G	H
	Dataset	
	14	
	12	
	15	
	16	
	11	
	19	
	13	
	12	
	16	
	15	
MIN	=MIN(G2:G11	
	MIN(**number1**, [number2], …)	

F	G	H
	Dataset	
	14	
	12	
	15	
	16	
	11	
	19	
	13	
	12	
	16	
	15	
MIN	11	

Fig. 3.15 Applying the MIN function

I	J	K	L
	Dataset		
	14		
	12		
	15		
	16		
	11		
	19		
	13		
	12		
	16		
	15		
MAX	=MAX(J2:J11		
	MAX(**number1**, [number2], …)		

I	J	K
	Dataset	
	14	
	12	
	15	
	16	
	11	
	19	
	13	
	12	
	16	
	15	
MAX	19	

Fig. 3.16 Applying the MAX function

the item's price by the number of units sold. Once we obtain the revenue from the items individually, we would then add the results to get the total revenue. Using the "SUMPRODUCT" function, Excel directly performs the entire calculation in a single line (Figs. 3.17, 3.18 and 3.19).

Goods	Unit Price
Apple Pie	$ 4.00
Banana Bread	$ 2.00
Caramel Mousse	$ 1.00
Date Pudding	$ 3.00
Fudge Brownie	$ 2.00
Lemon Tart	$ 4.00

Fig. 3.17   Sales in a bakery

	C	D	E	F	G
Goods		Unit Price	Units Sold		
Apple Pie		$ 4.00	8		
Banana Bread		$ 2.00	12		
Caramel Mousse		$ 1.00	6		
Date Pudding		$ 3.00	7		
Fudge Brownie		$ 2.00	15		
Lemon Tart		$ 4.00	9		
Total Revenue			=SUMPRODUCT(D2:D7,E2:E7)		
			SUMPRODUCT(array1, [array2], [array3], [array4], ...)		

Fig. 3.18   Applying the SUMPRODUCT function

C	D	E
Goods	Unit Price	Units Sold
Apple Pie	$ 4.00	8
Banana Bread	$ 2.00	12
Caramel Mousse	$ 1.00	6
Date Pudding	$ 3.00	7
Fudge Brownie	$ 2.00	15
Lemon Tart	$ 4.00	9
Total Revenue		149

Fig. 3.19   Total revenue

3.4.4 Homework Problem

1. Katherine runs a local business selling coffee and desserts. She uses her sales data to make a monthly performance report, which helps her evaluate how well her business runs. The following table shows her dessert sales data for a particular month.

Item	Quantity	Unit price
Chocolate tart	115	4.25
Blueberry crumble	64	3.75
Cream roll	82	2.5
Jam roll	77	3
Pecan pie	86	5
Pumpkin pie	68	4.75
Strawberry cheesecake	125	6.5
Strawberry shortcake	87	4
Devil's food cake	247	5.5
Angel's food cake	108	4.25

a. Find the SUM of sales.
b. Find the total revenue using SUMPRODUCT.
c. Which items have the highest and the lowest sales? Demonstrate the use of the MAX and MIN functions.

3.5 Trigonometric Functions

It helps to know some basic trigonometric operations that Excel has to offer. This section will look at unit conversion for angles and the three most basic trigonometric functions.

3.5.1 Units: Radians and Degrees

Excel offers direct unit conversion from degrees to radians and vice versa. The functions used are "DEGREES" and "RADIANS". If we have an angle in degrees and wish to convert it to radians, we simply need to type in "= RADIANS(theta)", where theta will be replaced with the value of the angle in degrees. Similarly, when converting an angle in radians to an angle in degrees, we use "= DEGREES(theta)". This time, theta represents the angle value in radians (Figs. 3.20, 3.21 and 3.22).

B	C
Angle (in Degrees)	Angle (in Radians)
30	=RADIANS(B2
45	RADIANS(angle)
60	
90	
180	

B	C
Angle (in Degrees)	Angle (in Radians)
30	0.524
45	0.785
60	1.047
90	1.571
180	3.142

Fig. 3.20 Converting degrees to radians

Fig. 3.21 Converting radians to degrees

B	C	D
Angle (in Degrees)	Angle (in Radians)	Angle (Reconverted to Degrees)
30	0.524	=DEGREES(C2)
45	0.785	
60	1.047	
90	1.571	
180	3.142	

Fig. 3.22 Angle unit conversion

B	C	D
Angle (in Degrees)	Angle (in Radians)	Angle (Reconverted to Degrees)
30	0.524	30
45	0.785	45
60	1.047	60
90	1.571	90
180	3.142	180

3.5.2 SIN, COS, TAN

The sine, cosine, and tangent values can also be found by specific functions, "SIN", "COS", and "TAN", respectively. These functions use the radian value of the angle as input. So, if the angles are in degrees, convert them to radians first, or directly convert them within the primary function. The process of enclosing one function within another is called nesting (Figs. 3.23, 3.24 and 3.25).

After entering the formula once, drag it down to apply the same formula pattern to the rest of the cells, as discussed earlier. Notice how the function returns an undefined value for the tangent of 90°. As we already know, 90° represent a line asymptote to the tangent curve, meaning a line that the curve cannot touch.

G	H
Angle (in Degrees)	**SIN**
30	=SIN(RADIANS(G2))
45	
60	
90	
180	

G	H
Angle (in Degrees)	**SIN**
30	0.500
45	0.707
60	0.866
90	1.000
180	0.000

Fig. 3.23 Using the SIN function

G	H	I
Angle (in Degrees)	**SIN**	**COS**
30	0.500	=COS(RADIANS(G2))
45	0.707	0.707
60	0.866	0.500
90	1.000	0.000
180	0.000	-1.000

Fig. 3.24 Using the COS function

G	H	I	J
Angle (in Degrees)	**SIN**	**COS**	**TAN**
30	0.500	0.866	0.577
45	0.707	0.707	(RADIANS(G3))
60	0.866	0.500	1.732
90	1.000	0.000	##############
180	0.000	-1.000	0.000

Fig. 3.25 Using the TAN function

3.6 Measures of Location and Variation

In the introduction chapter, we have discussed the different measures of location and variation, such as mean, median, mode, quartiles, percentiles, variance and standard deviation. In this section, we will learn how to compute these parameters using Excel.

3.6.1 Mean, Median, Mode

Microsoft Excel includes elementary built-in functions for finding a discrete dataset's mean, median and mode. Figure 3.11 shows the monthly cupcake sales for a bakery. Notice how most values are similar, while some are significantly larger or smaller than the rest (outliers). We will be using this example to learn about the following few functions (Fig. 3.26).

To find the mean, we have to use the AVERAGE function. First, add a heading for "Mean Sales". Next, type a "=" symbol followed by "AVERAGE" and start a bracket. Inside the bracket, we have to add our data. Place the cursor on the first cell containing the sales value for January. Press and hold. Drag it down to the sales for December (Fig. 3.27).

After selecting all the values, release the cursor hold, and the final formula should be placed into the cell automatically. Add the closing bracket and press enter (Fig. 3.28).

Notice how the result displays six decimal places. However, we know that there cannot be a fraction of the number of cupcake sales. So, we will format the number using the "Number" option above the worksheet (Fig. 3.29).

Click on the small arrow in the corner to launch the "Format Cells" tab. Select "Number" as the category, change the number of decimal places to zero, and click OK (Figs. 3.30 and 3.31).

Similarly, we can find the mode and median of the cupcake sales using the "MODE.SNGL" and the "MEDIAN" functions, respectively. Figures 3.17, 3.18, 3.19 and 3.20 demonstrates these functions and their results (Figs. 3.32, 3.33, 3.34 and 3.35).

Fig. 3.26 Monthly cupcake sales for a bakery

Month	Cupcake Sales
January	437
February	452
March	398
April	287
May	503
July	362
August	362
September	329
October	654
November	597
December	830

Fig. 3.27 Using the AVERAGE function

Fig. 3.28 Result obtained from the analysis

3.6.2 Variance and Standard Deviation

We can use several methods to find the variance of a sample or population using Excel. We can use built-in functions for variance directly, or we can break down the steps such as finding the mean, the deviations from the mean, the standard deviation and then the variance (Figs. 3.36 and 3.37).

Once we find the variance, we can calculate the standard deviation using the square root function or directly apply a built-in function for standard deviation (Figs. 3.38, 3.39 and 3.40).

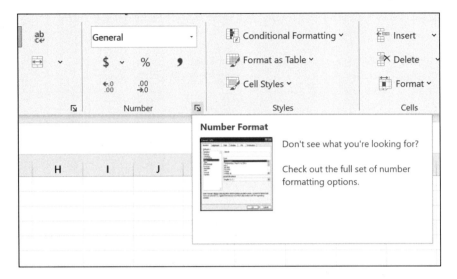

Fig. 3.29 Formatting number

Format Cells   ?   ✕

| Number | Alignment | Font | Border | Fill | Protection |

Category:

General
Number
Currency
Accounting
Date
Time
Percentage
Fraction
Scientific
Text
Special
Custom

Sample

  474

Decimal places: 0 ▲▼

☐ Use 1000 Separator (,)

Negative numbers:

-1234
1234
(1234)
(1234)

Number is used for general display of numbers. Currency and Accounting offer specialized formatting for monetary value.

OK   Cancel

Fig. 3.30 Changing the decimal places

Fig. 3.31 Final result for mean sales

	A	B	C
1			
2	Month	Cupcake Sales	Mean Sales
3	January	437	474
4	February	452	
5	March	398	
6	April	287	
7	May	503	
8	July	362	
9	August	362	
10	September	329	
11	October	654	
12	November	597	
13	December	830	

Fig. 3.32 Function for mode

	Month	Cupcake Sales	Mode
2			
3	January	437	=MODE.SNGL(
4	February	452	
5	March	398	
6	April	287	
7	May	503	
8	July	362	
9	August	362	
10	September	329	
11	October	654	
12	November	597	
13	December	830	

Fig. 3.33 Final result for mode

	Month	Cupcake Sales	Mode
2			
3	January	437	362
4	February	452	
5	March	398	
6	April	287	
7	May	503	
8	July	362	
9	August	362	
10	September	329	
11	October	654	
12	November	597	
13	December	830	

3.6.3 Quartiles and Percentiles

Like in the previous sections, we will also use built-in functions for quartiles and percentiles. To find quartiles, we have two options. Either we can include the

1			
2	**Month**	**Cupcake Sales**	**Median**
3	January	437	=MEDIAN(B3:B13
4	February	452	
5	March	398	
6	April	287	
7	May	503	
8	July	362	
9	August	362	
10	September	329	
11	October	654	
12	November	597	
13	December	830	

Fig. 3.34 Finding the median

1			
2	**Month**	**Cupcake Sales**	**Median**
3	January	437	437
4	February	452	
5	March	398	
6	April	287	
7	May	503	
8	July	362	
9	August	362	
10	September	329	
11	October	654	
12	November	597	
13	December	830	

Fig. 3.35 Final result for median

1			
2	**Month**	**Cupcake Sales**	**Variance**
3	January	437	=VAR.S(B3:B13
4	February	452	VAR.S(**number1**, [number2], …)
5	March	398	
6	April	287	
7	May	503	
8	July	362	
9	August	362	
10	September	329	
11	October	654	
12	November	597	
13	December	830	

Fig. 3.36 Finding variance

	A	B	C
1			
2	**Month**	**Cupcake Sales**	**Variance**
3	January	437	26382
4	February	452	
5	March	398	
6	April	287	
7	May	503	
8	July	362	
9	August	362	
10	September	329	
11	October	654	
12	November	597	
13	December	830	

Fig. 3.37 Result of variance

	A	B	C	D
1				
2	**Month**	**Cupcake Sales**	**Variance**	**Standard Deviation**
3	January	437	26382	=SQRT(C3)
4	February	452		
5	March	398		
6	April	287		
7	May	503		
8	July	362		
9	August	362		
10	September	329		
11	October	654		
12	November	597		
13	December	830		

Fig. 3.38 Standard deviation from variance

	A	B	C
1			
2	**Month**	**Cupcake Sales**	**Standard Deviation**
3	January	437	=STDEV.S(B3:B13
4	February	452	STDEV.S(**number1**, [number2], ...)
5	March	398	
6	April	287	
7	May	503	
8	July	362	
9	August	362	
10	September	329	
11	October	654	
12	November	597	
13	December	830	

Fig. 3.39 Standard deviation using built-in function

	Month	Cupcake Sales	Standard Deviation
2			
3	January	437	162
4	February	452	
5	March	398	
6	April	287	
7	May	503	
8	July	362	
9	August	362	
10	September	329	
11	October	654	
12	November	597	
13	December	830	

Fig. 3.40 Result of standard deviation

minimum and maximum values or exclude them. We use the "QUARTILE.INC" function for the first case, and for the second case, we use "QUARTILE. EXC". We must provide the data and specify the required quartile in both cases.

Say we want to find the first quartile of our dataset using the QUARTILE.INC function. The following figure illustrates the function in use (Fig. 3.41).

First, type in the function name and add the data. Then insert a comma and add the position of the quartile. Here, 1 represents the first quartile. Observe how Excel shows us what digit corresponds to the other quartiles, the maximum and the minimum value of the dataset. This numbering applies to the QUARTILE.INC function only. After entering the command, press enter. Figure 3.27 shows the results obtained for all five values (Fig. 3.42).

Similarly, the QUARTILE.EXC function can be used to find the first, second, and third quartiles, as shown in Fig. 3.28. Notice how this function has no option for finding the minimum or maximum value (Fig. 3.43).

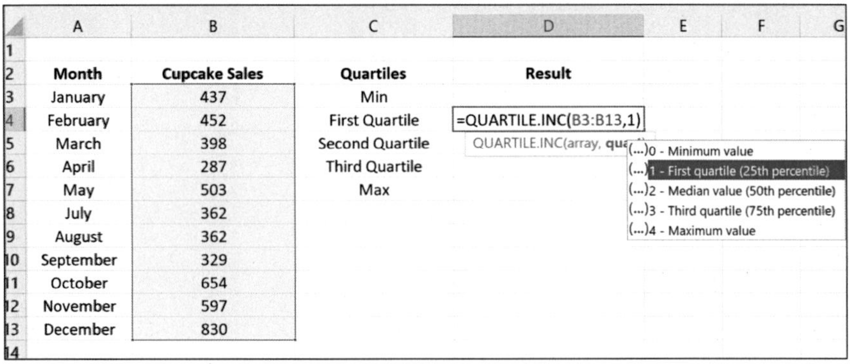

Fig. 3.41 Using QUARTILE.INC

	A	B	C	D
1				
2	**Month**	**Cupcake Sales**	**Quartiles**	**Result**
3	January	437	Min	362
4	February	452	First Quartile	287
5	March	398	Second Quartile	425
6	April	287	Third Quartile	597
7	May	503	Max	830
8	July	362		
9	August	362		
10	September	329		
11	October	654		
12	November	597		
13	December	830		
14				

Fig. 3.42 Results using QUARTILE.INC

	A	B	C	D	E	F
1						
2	**Month**	**Cupcake Sales**	**Quartiles**	**Result**		
3	January	437				
4	February	452	First Quartile	=QUARTILE.EXC(B3:B13,1)		
5	March	398	Second Quartile	QUARTILE.EXC(array, qua(...) 1 - First quartile (25th percentile)		
6	April	287	Third Quartile	(...)2 - Median value (50th percentile)		
7	May	503		(...)3 - Third quartile (75th percentile)		
8	July	362				
9	August	362				
10	September	329				
11	October	654				
12	November	597				
13	December	830				

Fig. 3.43 Using QUARTILE.EXC

The dataset's percentiles can also be found similarly, using the PERCENTILE.INC and PERCENTILE.EXC functions. This time, the number corresponding to a percentile will range from zero to one, inclusive, for PERCENTILE.INC and from zero to one, exclusive, for PERCENTILE.EXC. In Fig. 3.29, the 35th percentile is found using the exclusive function; in Fig. 3.30, the 0th percentile or the minimum value is found using the inclusive function (Figs. 3.44 and 3.45).

3.6.4 Box-and-Whisker Diagram

We have already been introduced to the box-and-whisker diagram in the introductory section. Now we will see how to plot such a graph using Excel. It is straightforward. Like the manual method, we need to know 5 points along the dataset—the median, the first and third quartiles, the maximum and the minimum values. Let us plot the graph with the dataset used in the previous section (Fig. 3.46).

	A	B	C	D
1				
2	**Month**	**Cupcake Sales**	**Quartiles**	**Result**
3	January	437		
4	February	452	35th Percentile	=PERCENTILE.EXC(B3:B13,0.35)
5	March	398	55th Percentile	
6	April	287	78th Percentile	
7	May	503		
8	July	362		
9	August	362		
10	September	329		
11	October	654		
12	November	597		
13	December	830		

Fig. 3.44 Using PERCENTILE.EXC

	A	B	C	D
1				
2	**Month**	**Cupcake Sales**	**Quartiles**	**Result**
3	January	437	0th Percentile	=PERCENTILE.INC(B3:B13,0)
4	February	452	35th Percentile	PERCENTILE.INC(array, **k**)
5	March	398	55th Percentile	
6	April	287	78th Percentile	
7	May	503	100th Percentile	
8	July	362		
9	August	362		
10	September	329		
11	October	654		
12	November	597		
13	December	830		

Fig. 3.45 Using PERCENTILE.INC

Fig. 3.46 Five points for
box-and-whisker plot

Minimum	362
1st Quartile	287
Median	425
3rd Quartile	597
Maximum	830

Select the five points and go to the "Recommended Charts" option from the Insert
tab. Select the option for box and whisker from "All Charts". Add the data labels
(Figs. 3.47, 3.48, 3.49, 3.50 and 3.51).

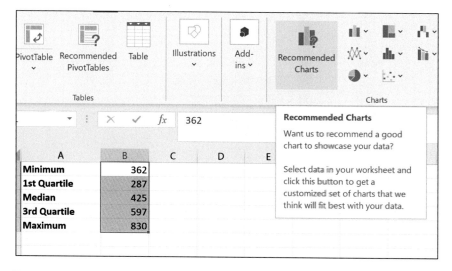

Fig. 3.47 Inserting chart process

Fig. 3.48 Selecting box-and-whisker plot

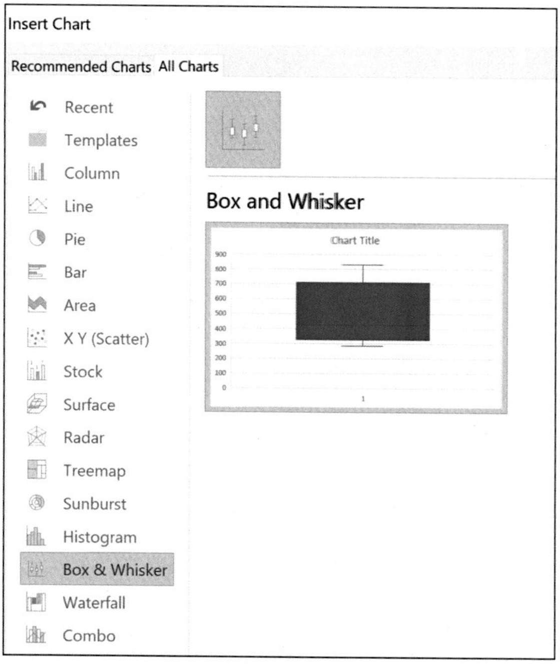

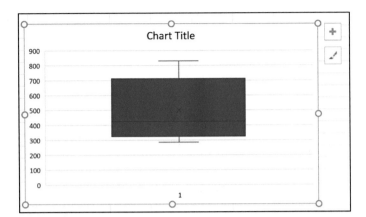

Fig. 3.49 Box-and-whisker plot obtained initially

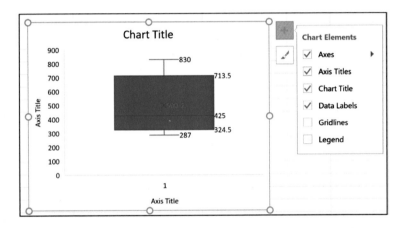

Fig. 3.50 Adding data labels

Right-click on the blue box and select "Format Data Series". Choose the option for "Inclusive Median". Delete the 500.2 value in the graph. The plot is essentially complete at this point. It can then be formatted as per requirements.

3.6.5   *Homework Problem*

1. The following table shows the list of subjects offered by a school for Ordinary (O') Level Examinations and the number of students registered for each subject.

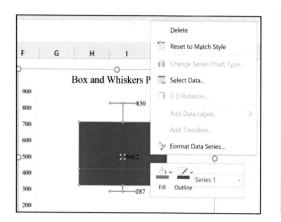 

Fig. 3.51 Formatting the plot

Subject	Students
Physics	215
Chemistry	200
Biology	198
Human biology	137
Mathematics	350
Advanced mathematics	295
Accounting	195
Economics	206
Commerce	203
Business studies	200
ICT	98

a. Find the average number of students per subject.
b. What is the median for this dataset?
c. Which subject has the highest number of students registered? Which subject has the lowest?
d. Find the 25th, 60th, 75th, and 90th percentile.
e. Draw a box plot using the 5-point summary for this dataset.

3.7 Correlation and Regression Modeling

When dealing with a large dataset, it is often vital to know the relationship between different parameters to predict the behavior of one variable with respect to changes in another. Correlation coefficients may be used to determine how strongly two variables are related. It can also tell us if they are positively or negatively related.

3.7.1 Pearson's Correlation Coefficient

Look at the dataset in Fig. 3.52. Looking at it, one can tell that Y increases when X increases. But, by how much? How strong is the response? A correlation coefficient value of zero indicates no correlation, whereas 1 indicates a perfect correlation. Let us try using the "CORREL" function first. Select the X values first, add a comma, and then select the Y values. Close the bracket and press enter (Figs. 3.53, 3.54 and 3.55).

This result shows that X and Y strongly correlate with a positive relationship. If X is increased 2.5 times, Y would increase 0.992×2.5 times, i.e., by a factor of 2.48. We can also use the "PEARSON" function to find the correlation similarly (Figs. 3.56 and 3.57).

Fig. 3.52 Final box-and-whisker diagram

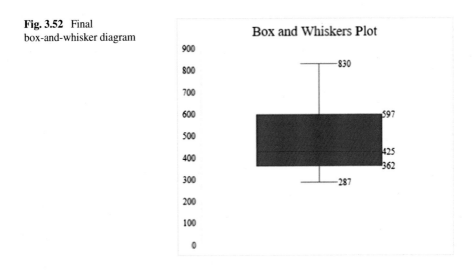

Fig. 3.53 Dataset

X	Y
10	125
20	268
30	343
40	526
50	600
60	789
70	863
80	1020
90	1235
100	1307

Fig. 3.54 Using CORRELATION function

	A	B	C
1	X	Y	Correlation Coefficient
2	10	125	=CORREL(A2:A11,B2:B11)
3	20	268	
4	30	343	
5	40	526	
6	50	600	
7	60	789	
8	70	863	
9	80	1020	
10	90	1235	
11	100	1307	
12			

Fig. 3.55 Result using the CORRELATION function

	A	B	C
1	X	Y	Correlation Coefficient
2	10	125	0.996
3	20	268	
4	30	343	
5	40	526	
6	50	600	
7	60	789	
8	70	863	
9	80	1020	
10	90	1235	
11	100	1307	

	A	B	C
1	X	Y	Correlation Coefficient
2	10	125	=PEARSON(A2:A11,B2:B11)
3	20	268	
4	30	343	
5	40	526	
6	50	600	
7	60	789	
8	70	863	
9	80	1020	
10	90	1235	
11	100	1307	

Fig. 3.56 Using the PEARSON function

C2			✕ ✓	fx	=PEARSON(A2:A11,B2:B11)

	A	B	C	D
1	X	Y	Correlation Coefficient	
2	10	125	0.996	
3	20	268		
4	30	343		
5	40	526		
6	50	600		
7	60	789		
8	70	863		
9	80	1020		
10	90	1235		
11	100	1307		

Fig. 3.57 Result using the PEARSON function

3.7.2 Scatter Diagrams and Trendlines

Taking the data in Fig. 3.52, we will plot a scatter diagram. Select the area containing the data. Go to the Insert Tab above the worksheet. From the charts section, select the option for a scatter diagram without any lines joining the points (Figs. 3.58 and 3.59).

Selecting that option will plot the points on a set of axes without any line connecting them. Now click on the option for chart elements and select trendline. We will also remove the gridlines, so unselect that option and add axis titles (Figs. 3.60 and 3.61).

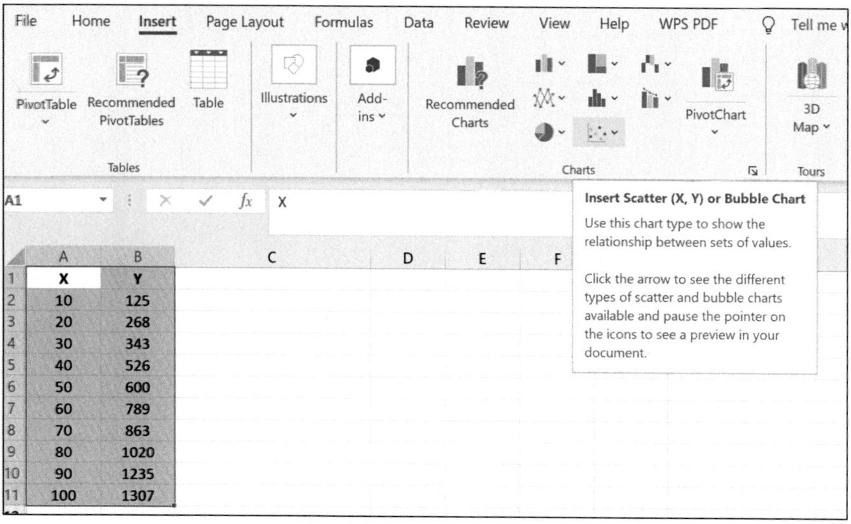

Fig. 3.58 Inserting chart steps

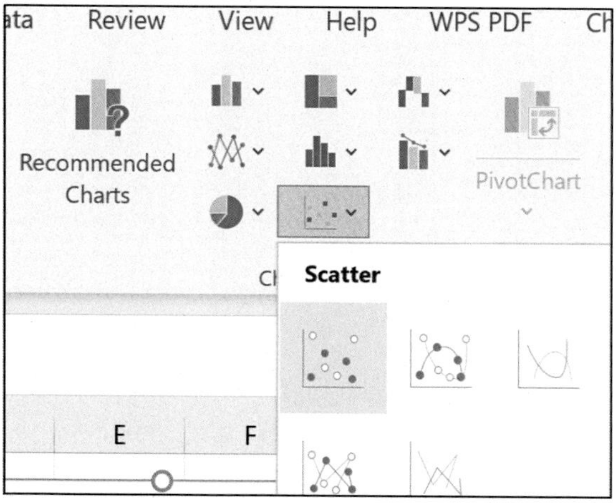

Fig. 3.59 Selecting scatter chart

The final result should look like this. The trendline is different from a line connecting the points. It tells us the relationship between the two variables, similarly to a correlation coefficient. We can see that the line does not go through the points, which means this is not a perfect correlation. However, the points lie close to the line, indicating a strong correlation. We can also tell from the positive slope that the parameters are positively linked.

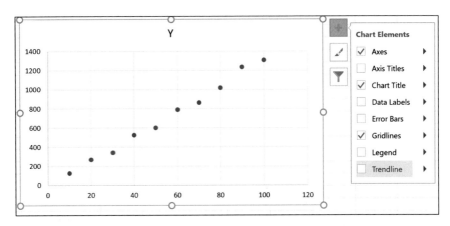

Fig. 3.60 Adding trendline

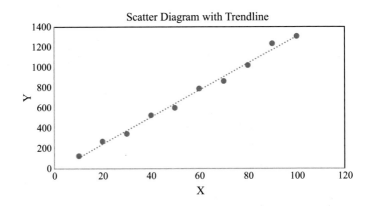

Fig. 3.61 Scatter diagram with trendline

3.7.3   Linear Models

So far, we have seen how to find a correlation between variables and plot the data points on a set of axes. We will learn to predict future values based on the present trend using two linear model functions.

3.7.3.1   FORECAST

Using the dataset in Fig. 3.52, suppose we want to predict a value of Y for a new value of X that is not included in the dataset. This function takes three arguments— the new X value, the known X values, and the known Y values. The function returns the predicted Y value for $X = 125$ (Figs. 3.62 and 3.63).

◢	A	B	C	D	E	F	G
1	X	Y					
2	10	125					
3	20	268		X	Forecasted Y value		
4	30	343		125	=FORECAST(D4,B2:B11,A2:A11		
5	40	526			FORECAST(x, known_y's, known_x's)		
6	50	600					
7	60	789					
8	70	863					
9	80	1020					
10	90	1235					
11	100	1307					

Fig. 3.62 Applying the FORECAST function

◢	A	B	C	D	E
1	X	Y			
2	10	125			
3	20	268		X	Forecasted Y value
4	30	343		125	1640.59
5	40	526			
6	50	600			
7	60	789			
8	70	863			
9	80	1020			
10	90	1235			
11	100	1307			

Fig. 3.63 Result using the FORECAST function

3.7.3.2 TREND

The "FORECAST" function returns one predicted value, but what if we wanted to find the predicted Y values for a whole range of X values? Here is where the "TREND" function comes into play. This function takes four arguments—the known Y values, the known X values, and the new X values, and the fourth argument can be left blank for regular computations. First, add the new X values to the worksheet. Then select the group of cells where the results should be displayed (Fig. 3.64).

	A	B	C	D	E	F	G
1	**X**	**Y**					
2	10	125					
3	20	268		**X**	**TREND of Y**		
4	30	343		125	=TREND(B2:B11,A2:A11,D4:D8)		
5	40	526		140	TREND(known_y's, [known_x's], [new_x's], [const])		
6	50	600		155			
7	60	789		170			
8	70	863		185			
9	80	1020					
10	90	1235					
11	100	1307					

Fig. 3.64 Using the TREND function

	A	B	C	D	E
1	**X**	**Y**			
2	10	125			
3	20	268		**X**	**TREND of Y**
4	30	343		125	1640.59
5	40	526		140	1840.52
6	50	600		155	2040.45
7	60	789		170	2240.38
8	70	863		185	2440.30
9	80	1020			
10	90	1235			
11	100	1307			

Fig. 3.65 Result using the TREND function

Start the function with "= TREND" and start a bracket. After entering the arguments or adding the data ranges, close the bracket and press control-shift-enter to complete the function (Fig. 3.65).

3.7.4 Nonlinear Models

The "FORECAST" and "TREND" functions are linear modeling functions. We will now explore a function for exponential functions.

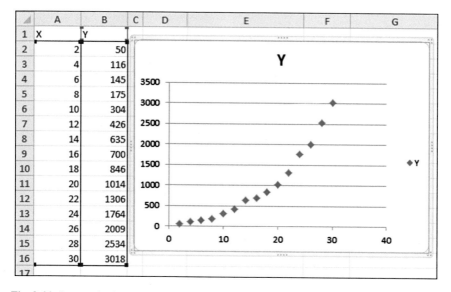

Fig. 3.66   Scatter plot for exponential dataset

3.7.4.1   GROWTH

Enter the data and plot a scatter diagram first to see the trend of the data. Then, like the "TREND" function, add the new × values to the worksheet. Select the group of cells where the results should be displayed. Enter the function and its arguments (Figs. 3.66 and 3.67).

After entering the arguments or adding the data ranges, close the bracket and press control-shift-enter to complete the function. Now let us add the new points to our scatter diagram. Right-click on the plot and click "Select Data" (Figs. 3.68 and 3.69).

The "Select Data Source" tab should open up. Select "Add". In the "Edit Series" tab, add the series name and select the X and Y values range (Figs. 3.70, 3.71 and 3.72).

Notice how the predicted values continue the curve for the existing values.

3.7.5   Homework Problem

1.1. The Uchiha fitness club has launched a new weight loss program for clients who have obesity problems. The following table shows the results obtained for a group of female clients between the ages of 25–40.

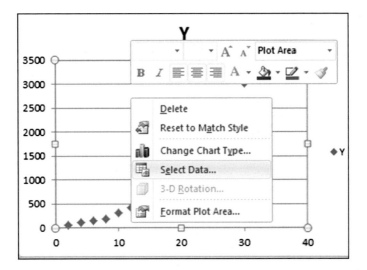

Fig. 3.67 Using the GROWTH function

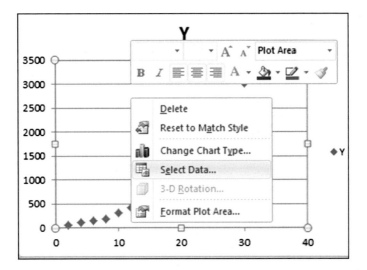

Fig. 3.68 Formatting the plot steps

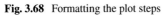

Month	Average weight (kg)
January	217
February	209

(continued)

(continued)

Month	Average weight (kg)
March	189
April	178
May	154
June	146
July	133
August	122
September	105
October	98
November	90
December	87

a. Draw a scatter diagram representing the progress for the year.
b. Assume an exponential trend and predict the average weight for the next 4 months.
c. Add the new points to the scatter plot.
d. Comment on the effectiveness of the program.

3.8 Graphs and Charts

Graphical representation is one of the most vital parts of data analysis, as we have already discussed in earlier sections. In this section, we will learn how to draw pie charts, bar charts, histograms, and stem-and-leaf diagrams in Excel.

3.8.1 Pie Charts

Pie charts in Excel are far simpler than they are in MATLAB. We will be using the same example used in the MATLAB section. Let us write our data in the worksheet first (Fig. 3.73).

Select the entire area containing the data. Go to Insert and notice there will be a section for charts. Click on the small arrow next to the pie chart to see all options. Select the desired style. A simple 2D graph should look something like Fig. 3.74.

Double-click on the chart to format it. The color of each segment can be changed individually, modify the title, font, legends, etc., or choose a specific style from the built-in options. We can even make a doughnut or a 3D pie chart (Fig. 3.75).

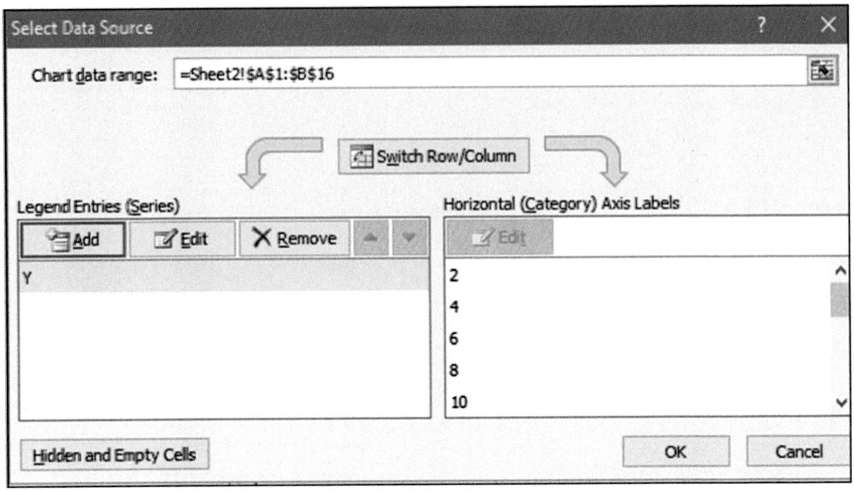

Fig. 3.69 Adding new series

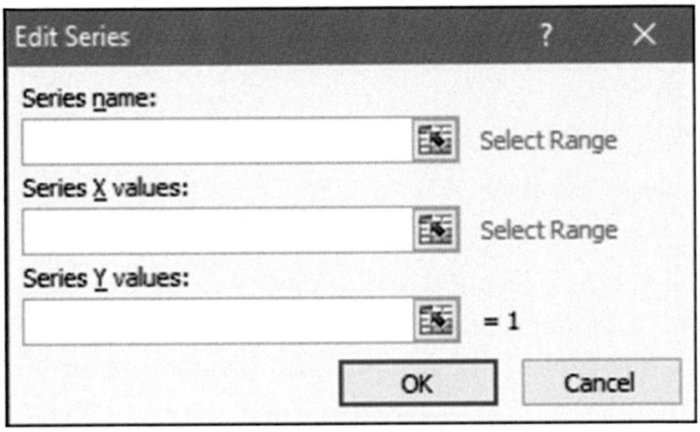

Fig. 3.70 Edit series tab

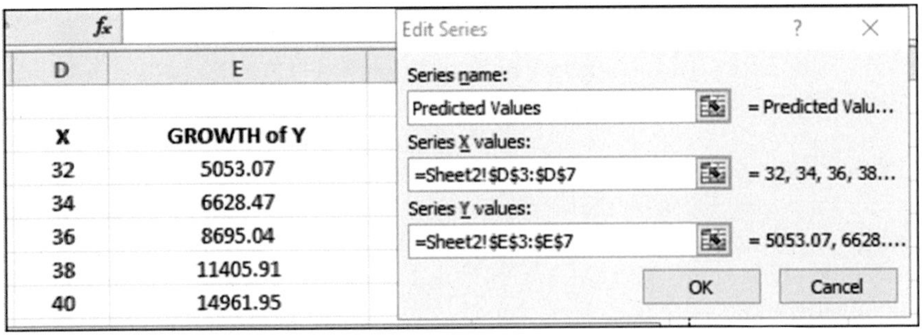

Fig. 3.71 Relevant information added to "Edit Series" tab

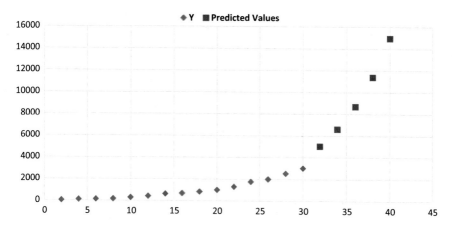

Fig. 3.72   Final scatter plot

Flavour	Number of Cupcakes
Chocolate	432
Strawberry	235
Orange	184
Vanilla	308

Fig. 3.73   Number of cupcakes of each flavor

3.8.2   Bar Charts

Bar charts are another effortless yet beneficial way to represent data, and Excel makes the process yet simpler with the option to generate bar charts instantly. Continuing with the dataset in Fig. 3.72, select the data and insert a column chart this time. Select the desired style and label or format the column chart obtained (Figs. 3.76 and 3.77).

Let us now try a horizontal bar chart using a different dataset. The process is the same as with the vertical column chart. Just select a 2D bar chart instead of a column chart this time (Fig. 3.78).

3.8.3   Histograms

Histograms in Excel take a little bit more work than bar charts. However, it is still a straightforward and quick process. Figure 3.79 shows the dataset we will be using to plot our histogram.

Fig. 3.74 Inserting pie chart

Fig. 3.75 Pie chart

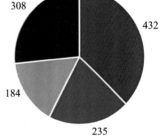

Like the previous charts, start by selecting the data and opening the Insert Tab. Select the "Insert Statistic Chart" option from the Charts section and click on "Histogram". Figure 3.80 shows the expected result (Fig. 3.81).

Select the x-axis, right-click on it, and then click the "Format Axis" option. A separate tab will open. Change the selection from "Automatic" to "By Category". The graph should then contain seven classes, each 10 kg in width (Fig. 3.82).

Once this is complete, it can be edited as needed. Figure 3.83 shows the final histogram plot. We used a built-in texture to fill the bars and adjusted the title and font. Notice how the gridlines may also be removed for a cleaner look.

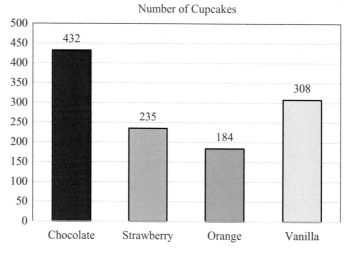

Fig. 3.76 Inserting column chart

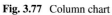

Fig. 3.77 Column chart

Fig. 3.78 Tiles sales

Tile	Sales
Tile A	249
Tile B	385
Tile C	320
Tile D	207

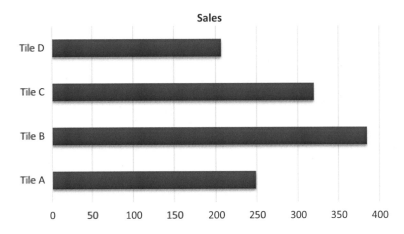

Fig. 3.79 Horizontal bar chart

Weight (kg)	No. of patients
30-40	8
40-50	15
50-60	27
60-70	32
70-80	16
80-90	6
90-100	12

Fig. 3.80 Weight of patients

3.8.4 Homework Problems

1. Represent the following data in a pie chart and a bar chart.

Color	Percentage
Blue	15
Green	25
Yellow	15
Red	45

2. At a college seminar, many students failed to show up on time. The following table shows a record of how late the students were. Draw a histogram representing the time delay.

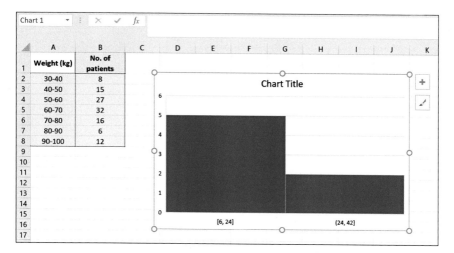

Fig. 3.81 Histogram generated initially

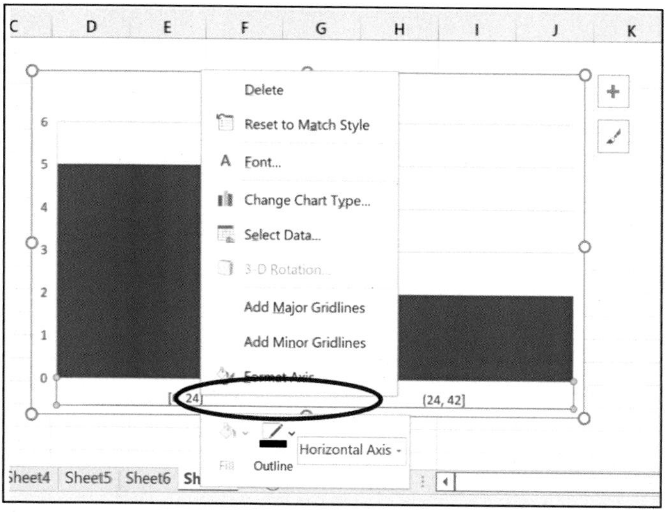

Fig. 3.82 Formatting axis

Delayed by (min)	No. of students
10–15	12
15–20	9

(continued)

(continued)

Delayed by (min)	No. of students
20–25	6
25–20	1
20–25	2

3.9   Financial Data Analysis

Financial data analysis is, without a doubt, an essential and vast subject. We will only be looking at some elementary functions, but the reader is encouraged to explore more finance or accounting-based functions that Excel offers.

3.9.1   Currency Formatting

Excel provides specialized format and symbols to represent currencies and financial data. Figure 3.84 shows the monthly revenue for an event management company. A dollar ($) symbol can be seen in the Number tab above the worksheet. First, select the data. Next, click on the $ symbol, and select "More Accounting Formats". A new "Format Cells" tab should appear. The required currency and decimal places may be selected from there (Figs. 3.85 and 3.86).

After applying the format changes, the result should look like this. Any negative number would be represented within brackets, per the accounting format (Fig. 3.87).

3.9.2   Simple and Compound Interest

Generally, loan repayment is made in periodic installments. In most cases, the payment is made on a monthly basis. Each month, there is an additional cost attached to the actual loan. This extra amount is the cost of acquiring the loan, known as the interest.

Interest is charged as a percentage of the total loan. There are two basic methods of charging interest: the simple interest method and the compound interest method. We will be looking at the two methods using an example.

Let us assume a $5000 loan was taken for 4 years at an interest rate of 5% per annum. What will be the future value of this loan at the end of the 4 years?

Using simple interest, we do not need a function to find the future value. Enter the present value, interest rate and payment period in Excel. The total interest accumulated at the end of the payment period can be found by multiplying the three terms.

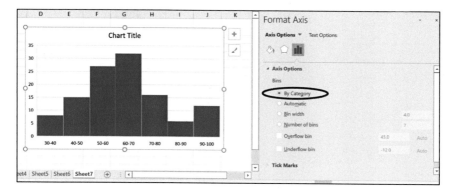

Fig. 3.83 Choosing "By Category"

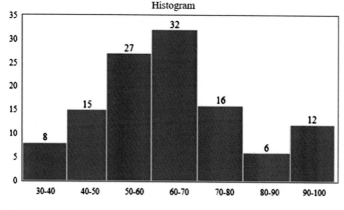

Fig. 3.84 Final histogram

Then, the repayment value would be the summation of the present value and the interest rate (Figs. 3.88 and 3.89).

Now let us solve the problem using compound interest. Enter the present value of the loan as a negative figure here, and the accounting format should put it within brackets. This represents a cash outflow. We will be using the "FV" function (Fig. 3.90).

The monthly interest rate is found by dividing the 5% by 12. The "nper" denotes the number of periods, which can be found by multiplying the number of years and the number of months each year. We are not adding periodic payments here, so the "pmt" here will be zero. Finally, the present value is added for the "pv" term. We do not need to enter any argument for the "type". The function returns the future value directly. If we want to know the accumulated interest, we can subtract the present value from the future value (Fig. 3.91).

Fig. 3.85 Currency
formatting

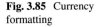

	$	✓	%	﹐		Fon
		$ English (United States)			ell	
	⤢	£ English (United Kingdom)				
		€ Euro (€ 123)				
		¥ Chinese (China)				
		CHF French (Switzerland)			M	
		More Accounting Formats...				

Month	Revenue
January	43700
February	45205
March	39847
April	28743
May	50300
July	36205
August	36224
September	32997
October	65460
November	59700
December	83700

3.9.3 Number of Payments (NPER & PDURATION)

As we discussed earlier, loan repayment is made in monthly installments. A payment plan can help us stay on track and find an optimum loan repayment solution. We will now look at two options provided by Excel, which can help us determine the repayment duration.

The first option is using the NPER function. We need to know three things before applying this function. First, how much is the principal amount? Next, what is the annual interest rate? Finally, how much can we pay monthly? Once we know the value of these three arguments, we can apply the function as illustrated in Fig. 3.92.

Here, a loan of $5000 was taken at an interest rate of 5% per annum. The interest rate of 0.05 is for a whole year. So, if we want the monthly interest rate, we must divide the value by 12. The monthly payment is written with a negative sign to indicate a cash outflow (Fig. 3.93).

After entering the values, select the monetary values and format them using the standard accounting and currency format. Notice how a set of brackets represents the negative value. Enter the function and the arguments and hit enter (Fig. 3.94).

The result 10.24 indicates that if we pay $500 every month, we will be able to complete our entire loan repayment, including the interest, within 10.24 months, i.e., a little over 10 months or with 11 payments.

Similarly, we can use the PDURATION function to find the repayment duration or the number of payments. However, the input values required are slightly different.

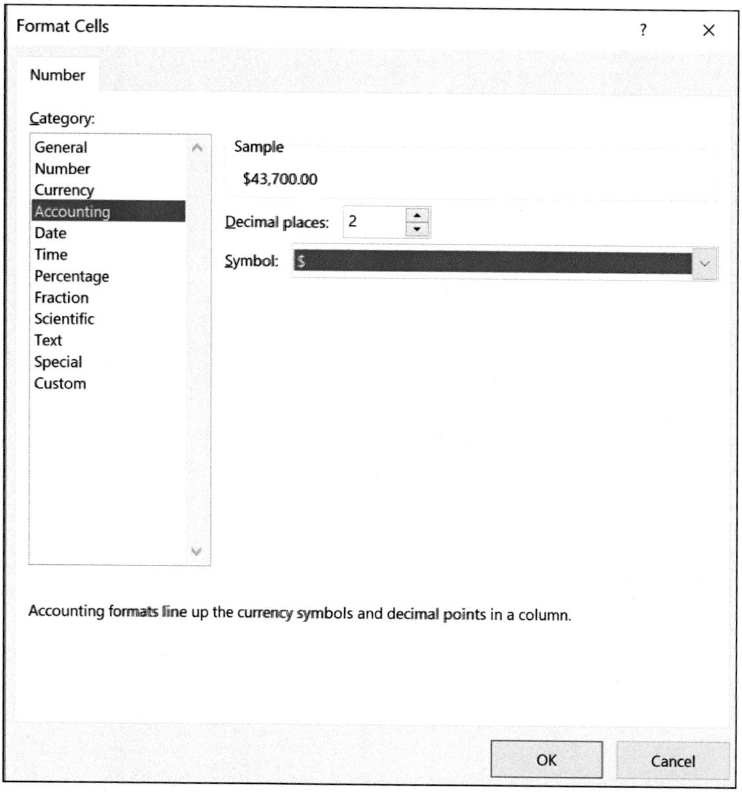

Fig. 3.86 Selecting currency and decimal places

Fig. 3.87 Formatted
revenue

Month	Revenue
January	$ 43,700.00
February	$ 45,205.00
March	$ 39,847.00
April	$ 28,743.00
May	$ 50,300.00
July	$ 36,205.00
August	$ 36,224.00
September	$ 32,997.00
October	$ 65,460.00
November	$ 59,700.00
December	$ 83,700.00

	A	B
1	Present Value	$ 5,000.00
2	Interest Rate	0.05
3	Number of Years	4
4	Future Value	=B1+(B1*B2*B3)
5		

Fig. 3.88 Simple interest

	A	B
1	Present Value	$ 5,000.00
2	Interest Rate	0.05
3	Number of Years	4
4	Future Value	$ 6,000.00
5		

Fig. 3.89 Future value using simple interest

	A	B	C
1	Present Value	$ (5,000.00)	
2	Interest Rate	0.05	
3	Number of Years	4	
4	Months per year	12	
5	Future Value	=FV(B2/12,B3*B4,0,B1)	
6		FV(rate, nper, pmt, **[pv]**, [type])	
7			

Fig. 3.90 Compound interest

We need the principal amount and the annual interest rate as before. Instead of the monthly payment, we need the total amount paid by the end of the payment period. This is useful when we are unsure how much we should pay every month, but we know the lump sum that must be paid for the entire loan (Fig. 3.95).

Enter the required items in the worksheet first. The total payment amount must not be written with a negative sign for the PDURATION function. Format the values and input the function (Figs. 3.96 and 3.97).

	A	B	C
1	Present Value	$ (5,000.00)	
2	Interest Rate	0.05	
3	Number of Years	4	
4	Months per year	12	
5	Future Value	$6,104.48	
6			

Fig. 3.91 Future value using compound interest

Fig. 3.92 Loan details

Principal Amount	$	5,000.00
Interest Rate		0.05
Monthly Payment	$	500.00

Fig. 3.93 Using NPER function

	A	B	C
1	Principal Amount	$ 5,000.00	
2	Interest Rate	0.05	
3	Monthly Payment	$ (500.00)	
4			
5	Number of Payments	=NPER(B2/12,B3,B1)	
6		NPER(rate, pmt, pv, [fv], [type])	
7			

Fig. 3.94 Number of payments

	A	B
1	Principal Amount	$ 5,000.00
2	Interest Rate	0.05
3	Monthly Payment	$ (500.00)
4		
5	Number of Payments	10.24

Fig. 3.95 Loan details in the table

Principal Amount	$	5,000.00
Interest Rate		0.05
Total Amount to be Paid	$	5,500.00

Fig. 3.96 Using PDURATION function

▲	A	B	C
1	Principal Amount	$ 5,000.00	
2	Interest Rate	0.05	
3	Total Amount to be Paid	$ 5,500.00	
4			
5	Number of Payments	=PDURATION(B2/12,B1,B3)	
6		PDURATION(**rate**, pv, fv)	

Fig. 3.97 Number of payments in the table

▲	A	B
1	Principal Amount	$ 5,000.00
2	Interest Rate	0.05
3	Total Amount to be Paid	$ 5,500.00
4		
5	Number of Payments	22.92
6		

From the result obtained, we can see that 23 payments, or 23 months, would be required to complete the payment. If we divide the $5500 by 23, we can determine the monthly amount to be paid.

3.9.4 Depreciation Using the Straight-Line Method

Depreciation is the declining value of fixed assets over the years due to various factors such as time, wear and tear, and new technology. It is a vast topic in finance and accounting, so it is beyond the scope of this book to explore the details of calculating depreciation. However, we will look at a very basic method of computing depreciation, the straight-line method (Fig. 3.98).

Enter the asset's cost, the salvage or resale value, and the years the asset will be in use. Then use the "SLN" function to find the yearly depreciation (Fig. 3.99).

Fig. 3.98 Using SLN function

	A	B	C
1	Cost of Asset	$ 5,000.00	
2	Salvage Value	$ 2,000.00	
3	Number of Years	5	
4	Depreciation Per Year	=SLN(B1,B2,B3)	
5			

SLN(cost, salvage, **life**)

Fig. 3.99 Depreciation per year

	A	B
1	Cost of Asset	$ 5,000.00
2	Salvage Value	$ 2,000.00
3	Number of Years	5
4	Depreciation Per Year	$600.00
5		

3.9.5 Homework Problems

1. The following table shows the details of a $10,000 loan taken at an interest rate of 4% per annum for 5 years.

Principal amount	$10,000
Annual interest rate	4%
No. of years	5

a. Calculate the total amount paid at the end of the 5 years, using simple interest.
b. Calculate the total amount paid at the end of the 5 years using compound interest. What is the accumulated compound interest at the end of the period?

2. The following table shows the details of a $6000 loan taken at an interest rate of 3% per annum. Calculate the number of payments needed to complete the loan repayment.

Principal amount	$6000
Annual interest rate	3%
Monthly payment	$500

3. The following table shows the details of a $3000 loan taken at an interest rate of 2% per annum. At the end of the payment period, the loan's value is $3120. Calculate the number of payments needed to complete the loan repayment.

Principal amount	$3000
Annual interest rate	2%
Future value	3120

4. Simon buys a machine for $7500 for his factory. After 5 years of service, he decides to sell off this machine and buy a new one. He finds out that the resale value for the machine will be $2000. What is his yearly depreciation on a straight-line basis?

3.10 Exercises

1. Tim purchases a lab apparatus worth $3000. He estimates that he will be using it for 3 years. Once his project is complete, he intends to sell it off. Assuming that the resale value for the machine will be $1000, what is his yearly depreciation using the straight-line method?
2. A wholesale stationery seller keeps a record of all the items sold every day. The following table shows the different articles and the quantities in which they were sold over a particular month.

Item	No. of boxes	Quantity per box
Pencils	128	12
Erasers	36	10
Sharpeners	72	6
Rulers	207	10
Sign pens	131	15
Markers	164	10
Compass	120	12
Pens	250	20
Whitening ink	35	12

a. Find the SUM of sales.
b. Find the total number of articles sold using SUMPRODUCT.
c. Which items have the highest and the lowest sales? Demonstrate the use of the MAX and MIN functions.
3. The following table shows the monthly sales of sugar from a departmental store.

Month	Sales (kg)
January	77
February	80
March	88
April	95
May	112
June	123
July	136
August	144
September	168
October	179
November	199
December	207

a. Draw a scatter diagram representing the sales trend.
b. Assume an exponential trend and predict the average weight for the next 3 months.
c. Add the new points to the scatter plot and comment on the trend.
4. The following table shows the percentage of students playing different sports hosted by a school club. Represent the following data in a pie chart and a bar chart.

Sport	Percentage
Volleyball	10
Cricket	30
Badminton	20
Football	40

5. At a merchandise sale, customers from different age groups attended. The following table shows the customer records. Draw a histogram representing the age distribution.

Age group (years)	No. of customers
20–25	7
25–30	8
30–35	6
35–40	4
40–45	3

6. The following table shows the score obtained by patients in a mental health test.

Patient serial	Students
01	115
02	100
03	98
04	87
05	110
06	125
07	95
08	106
09	103
010	100
11	98

a. Find the average score for the test.
b. What is the median for this dataset?
c. Which patient has the highest score? What is the lowest score? Apply functions in Excel to find the answers.
d. Find the 35th, 70th, 85th, and 90th percentile.
e. Draw a box plot using the 5-point summary for this dataset.

Chapter 4
SPSS

Abstract This chapter of "Statistics and Data Analysis for Engineers and Scientists" immerses readers in the powerful realm of SPSS, a statistical software package widely used for data analysis in various fields. This chapter provides a comprehensive introduction to SPSS, guiding readers through the essential steps for creating and defining variables, entering and sorting data. Building on this foundation, the chapter delves into measures of dispersion, elucidating concepts of central tendency and dispersion, quartiles, percentiles, and box-and-whisker plots. A detailed exploration of frequency distribution further enhances readers' understanding of data variability. The chapter also unfolds the intricate world of correlation analysis, employing scatter plots and Pearson's correlation coefficient to measure relationships between variables. Readers gain proficiency in regression analysis, particularly simple linear regression, enabling them to model and interpret linear relationships within data. Visualization capabilities are expanded with guidance on creating line graphs, coding data, crafting bar charts, pie charts, and histograms. The chapter culminates with an exploration of inferential statistics, focusing on the analysis of variance (ANOVA) technique to assess differences between group means. Exercises peppered throughout the chapter empower readers to apply newfound knowledge and hone their SPSS skills. By the chapter's end, readers will have unlocked SPSS's potential as a robust tool for data analysis, correlation and regression modeling, visualization, and inferential statistics. They will be well-equipped to tackle complex engineering and scientific challenges, making informed decisions grounded in rigorous statistical analysis.

Keywords Introduction to SPSS · Creating and defining variables · Entering and sorting data · Measures of dispersion · Central tendency and dispersion · Quartiles and percentiles · Box-and-whisker plot · Frequency distribution · Correlation · Scatter plots · Pearson's correlation coefficient · Regression analysis · Simple linear regression · Graphs and charts · Analysis of variance (ANOVA)

4.1 Introduction to SPSS

Just like MATLAB and Excel, SPSS is an excellent software for statistics and data analysis. Unlike Excel, SPSS provides all built-in functions for simple and complex analysis procedures. The interface is similar to Excel, so it should be easy to learn if one has some knowledge of Excel. It is also relatively simple when compared to MATLAB. Once we get a grip on how to use SPSS, an entire world of possibilities is waiting for us. Data Analysis can be performed very efficiently with only a few clicks using SPSS. There are options for computing dispersion or central tendency measures, such as mean, median, mode, variance, and standard deviation. Regression models, correlation coefficients, ANOVA, and other hypothesis tests can be designed or conducted. SPSS also offers a broad scope for the graphical representation of datasets and their nature.

For statistical analysis in social science, SPSS is a frequently used application. Government agencies, survey firms, marketing firms, market research groups, health researchers, data miners, and other entities use it. One of sociology's most significant publications, the original SPSS handbook, made it possible for common researchers to do their statistical analysis. Data administration and documentation are components of the basic program and statistical analysis. SPSS is an excellent program for statistics and data analysis, similar to MATLAB and Microsoft Excel. Unlike Excel, SPSS provides all built-in functions for simple and complex analysis procedures. The user interface is comparable to Excel, so it should be simple to learn for those who are familiar with Excel. It is also relatively simple when compared to MATLAB. Once we have mastered how to utilize SPSS, a universe of opportunities awaits us. Data analysis can be performed very efficiently with only a few clicks using SPSS. There are options for computing dispersion or central tendency measures, such as mean, median, mode, variance, and standard deviation. Regression models, correlation coefficients, ANOVA, and other hypothesis tests can be designed or conducted. SPSS also offers a broad scope for the graphical representation of datasets and their nature (Fig. 4.1).

4.1.1 Creating and Defining Variables

The very first step to working with SPSS is defining the variables involved. First, let us take a look at the interface. There are two different views—the Data View and the Variable View. We have to define our variables and their properties in the Variable View. The dataset is then entered in the Data View (Fig. 4.2).

The following table shows information related to students in Hyuga High School. Hiashi Hyuga, the headmaster, wants to learn about the overall performance of the students. He also wishes to find out if there is any prominent relationship between the different parameters.

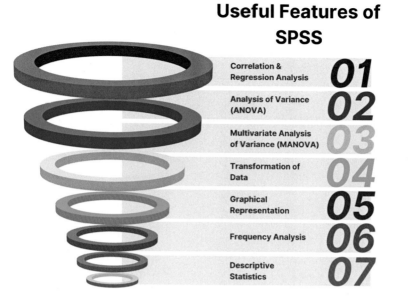

Useful Features of SPSS

Correlation & Regression Analysis **01**

Analysis of Variance (ANOVA) **02**

Multivariate Analysis of Variance (MANOVA) **03**

Transformation of Data **04**

Graphical Representation **05**

Frequency Analysis **06**

Descriptive Statistics **07**

Fig. 4.1. Useful features of SPSS

Fig. 4.2 Interface of SPSS

Student ID	Name	Physics score	Mathematics score
001	Alex	92	95
002	Rimon	90	100
003	Tanvir	100	97
004	Rashed	85	83
005	Shehrin	78	86
006	Ahmed	89	96
007	Niloy	73	76
008	Ram	74	82
009	Riyad	76	87
010	Mehnaz	84	86
011	Naushin	91	89
012	Mohammad	84	88
013	Julia	53	50
014	Imtiaz	90	86
015	Mustafy	68	72
016	Tahmeed	83	80
017	Rokib	95	100
018	Ehsan	96	98
019	Eren	39	54
020	Loki	68	65
021	Farid	100	100
022	Mesbah	75	79
023	Dinesh	88	93
024	Mahmud	82	97
025	Shishir	56	59
026	Marie	81	94
027	Andy	74	71
028	Stuart	91	94
029	Fredrick	64	87
030	Flintoff	75	89

In the dataset above, we have the test scores of a group of students for two different subjects. Considering the identification number and name of the students, we have four variables in total. There are 11 different properties we can define for each variable.

- Name: This is the variable name displayed as the column heading. It is short and generally non-informative.
- Type: We can choose from nine different variable types, as shown in Fig. 4.3. We can use the numeric or the restricted numeric type for the identification number of students. The restricted numeric type ensures that only a fixed number of digits may be entered, and if the number of digits is lower than the maximum number

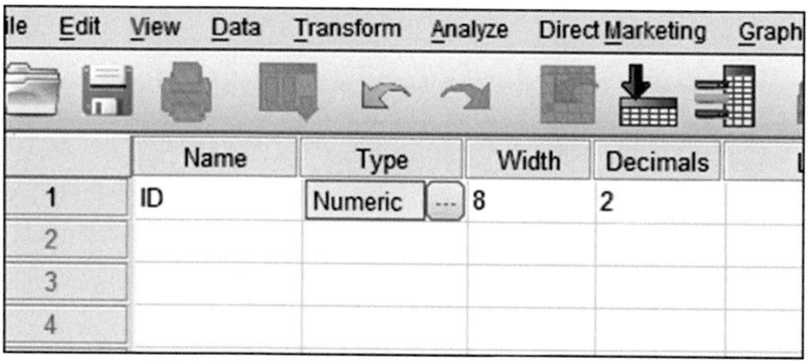

Fig. 4.3 Variable type

allowed, then the entry will be led with zeros. A string variable type is appropriate for the student names, whereas a numeric type suits the test scores.

- Width: This is the maximum number of characters that can be entered.
- Decimals: This defines the number of decimal places. For the example above, we do not need any decimal points for the ID number or the test scores. However, we have demonstrated using two decimal places for the test scores.
- Label: This option is provided to add a description for the variable, which we could not do in the Name section. It is more informative and helps the user or reader understand the variable better.
- Values: "Values" is a valuable property that allows us to code our data in terms of numbers, letters, words, etc. We will explore this property in a later section.
- Missing: This option can be used to define any value that we do not wish to include in the data analysis. We can also leave it empty, and SPSS will automatically interpret it as missing data.
- Columns: "Columns" defines the column width in the Data View. Generally, 8 is set as the default. We can either change this value or alter the width of the column directly in the Data View.
- Align: We can choose our desired alignment for the data.
- Measure: When dealing with names or categorical values, we use the Nominal measure. For rankings, we use the Ordinal measure. As for continuous data, we have to select Scale.
- Role: Select "Input" as the role since we will be inputting our data under these variables (Figs. 4.4 and 4.5).

4.1.2 Entering and Sorting Data

Figure 4.6 shows the Data View of SPSS. The main data analysis work is done here in the Data View once the data is entered under the variables that have been defined. As the figure shows, the defined variables have been added as the headings of columns. This heading stays fixed unless it is changed again in the Variable View. It means that one entire column is assigned to a single variable (Figs. 4.7 and 4.8).

Fig. 4.4 Defining variable type and width

	Name	Type	Width	Decimals	Label
1	ID	Restricted ...	3	0	Identification Number of the Student
2	Name	String	30	0	Name of the Student
3	Physics	Numeric	3	0	Score in Physics Test
4	Math	Numeric	3	0	Score in Mathematics Test
5					
6					
7					

Fig. 4.5 Variable labels

Name	Type	Width	Decimals	Label	Values	Missing	Columns	Align	Measure	Role
ID	Restricted ...	3	0	Identification N...	None	None	8	Left	Ordinal	Input
Name	String	30	0	Name of the St...	None	None	21	Left	Nominal	Input
Physics	Numeric	3	0	Score in Physi...	None	None	8	Right	Scale	Input
Math	Numeric	3	0	Score in Mathe...	None	None	8	Right	Scale	Input

Fig. 4.6 Defining variables

Next, enter the data. Notice how the ID numbers start with zeros. If we type 1, we will get 001. This format is a result of the restricted numeric type selected earlier. Since a width of 3 was chosen, all numbers less than three digits will automatically have zeros at the beginning. Also, note that there are two decimal places for the test scores. This is because we defined the test scores with two decimal places.

If we want to remove the digits after the decimal point, we need to modify the decimal property in the Variable View. The decimal places are not needed for this dataset, so it does not make a difference. However, we can freely add or reduce decimal points as needed. Go to the Variable View and change the "Decimals" property to 0 (Fig. 4.9).

Fig. 4.7 Data view

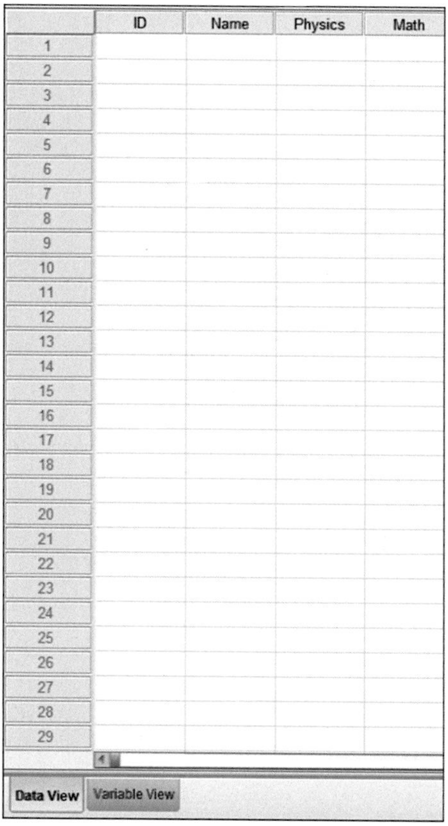

4.2 Measures of Dispersion

4.2.1 Central Tendency and Dispersion

In this section, we will use SPSS for primary descriptive statistical analyses. Go to the "Analyze" toolbar above the dataset. Look for the "Frequencies" option from the drop-down menu that appears. Select "Descriptive Statistics". A new window called "Frequencies" will pop up. Select the desired variable or variables. We will work with the "Score in Physics" variable in this case. Once the desired variable is selected, click on the arrow in the middle to add it to the variable list (Figs. 4.10 and 4.11).

In the same way, add the "Score in Mathematics" to the list as well. If any variable is added accidentally and needs to be removed, select it, and click on the arrow again. The arrow should be pointing in the opposite direction this time.

Click the "Statistics" option. A new window will appear, as shown in Fig. 4.12. Here, the desired boxes have to be ticked. Select the mean, mode, median, and sum

Fig. 4.8 Data entry with
decimal places

	ID	Name	Physics	Math
1	001	Alex	92.00	95.00
2	002	Rimon	90.00	100.00
3	003	Tanvir	100.00	97.00
4	004	Rashed	85.00	83.00
5	005	Shehrin	78.00	86.00
6	006	Ahmed	89.00	96.00
7	007	Niloy	73.00	76.00
8	008	Ram	74.00	82.00
9	009	Riyad	76.00	87.00
10	010	Mehnaz	84.00	86.00
11	011	Naushin	91.00	89.00
12	012	Mohammad	84.00	88.00
13	013	Julia	53.00	50.00
14	014	Imtiaz	90.00	86.00
15	015	Mustafy	68.00	72.00
16	016	Tahmeed	83.00	80.00
17	017	Rokib	95.00	100.00
18	018	Ehsan	96.00	98.00
19	019	Eren	39.00	54.00
20	020	Loki	68.00	65.00
21	021	Farid	100.00	100.00
22	022	Mesbah	75.00	79.00
23	023	Dinesh	88.00	93.00
24	024	Mahmud	82.00	97.00
25	025	Shishir	56.00	59.00
26	026	Marie	81.00	94.00
27	027	Andy	74.00	71.00
28	028	Stuart	91.00	94.00
29	029	Fredrick	64.00	87.00

Data View   Variable View

boxes under the "Central Tendency" section. Also, select the boxes for variance, standard deviation, range, minimum and maximum values. SPSS will provide direct answers to these. After selecting the required options, click Continue and then OK. SPSS will display the results in a tabular format. Figure 4.13 shows the results, which appear in a separate window (Fig. 4.14).

4.2.2   Quartiles and Percentiles

Quartiles and percentiles can be computed in the same way as the other measures of dispersion shown in the previous section. The only difference is that we need to specify which percentiles we want. Select the option for percentiles, write down the required value, and press "Add". Similarly, add any other percentiles desired. Click Continue and then OK. The results should appear in a new window (Figs. 4.15, 4.16 and 4.17).

Fig. 4.9 Removing the decimal places

	ID	Name	Physics	Math
1	001	Alex	92	95
2	002	Rimon	90	100
3	003	Tanvir	100	97
4	004	Rashed	85	83
5	005	Shehrin	78	86
6	006	Ahmed	89	96
7	007	Niloy	73	76
8	008	Ram	74	82
9	009	Riyad	76	87
10	010	Mehnaz	84	86
11	011	Naushin	91	89
12	012	Mohammad	84	88
13	013	Julia	53	50
14	014	Imtiaz	90	86
15	015	Mustafy	68	72
16	016	Tahmeed	83	80
17	017	Rokib	95	100
18	018	Ehsan	96	98
19	019	Eren	39	54
20	020	Loki	68	65
21	021	Farid	100	100
22	022	Mesbah	75	79
23	023	Dinesh	88	93
24	024	Mahmud	82	97
25	025	Shishir	56	59
26	026	Marie	81	94
27	027	Andy	74	71
28	028	Stuart	91	94
29	029	Fredrick	64	87

Data View Variable View

Fig. 4.10 Descriptive statistics

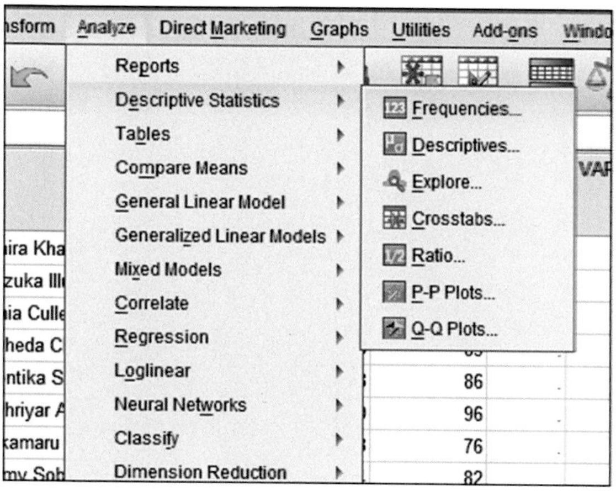

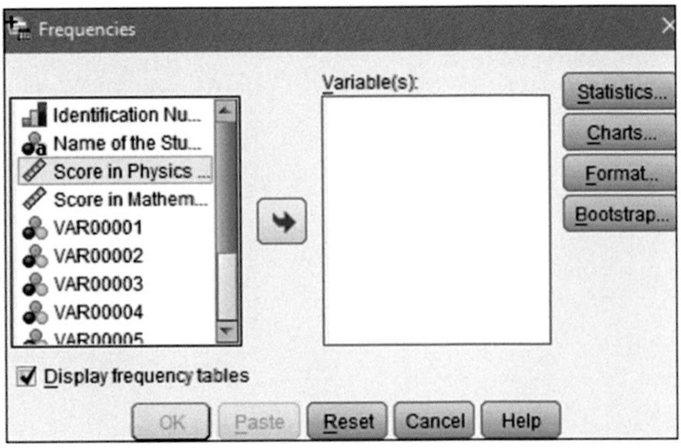

Fig. 4.11 Adding variables

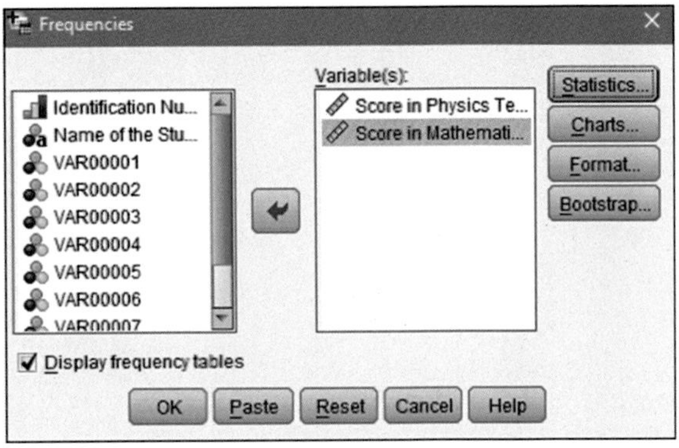

Fig. 4.12 Variables added
for analysis

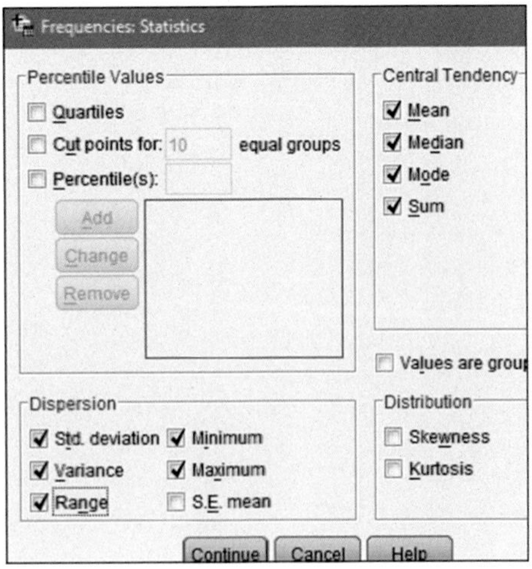

Fig. 4.13 Selecting required
options

→ Frequencies

[DataSet1] C:\Users\PCFIX\Desktop\Mahira\Data Analysis Book\SPSS\Untitled1.sav

Statistics

		Score in Physics Test	Score in Mathematics Test
N	Valid	30	30
	Missing	0	0
Mean		79.80	84.43
Median		82.50	87.00
Mode		68[a]	86[a]
Std. Deviation		14.146	13.650
Variance		200.097	186.323
Range		61	50
Minimum		39	50
Maximum		100	100
Sum		2394	2533

a. Multiple modes exist. The smallest value is shown

Fig. 4.14 Result table generated

Fig. 4.15 Adding percentiles

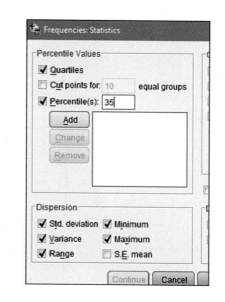

4.2.3 Box-and-Whisker Plot

Now that we have generated the mean, mode, median, standard deviation, quartiles, and percentiles, let's represent the nature of the dataset using a box plot. Click on the Graphs toolbar. From the drop-down menu that appears, select the Legacy

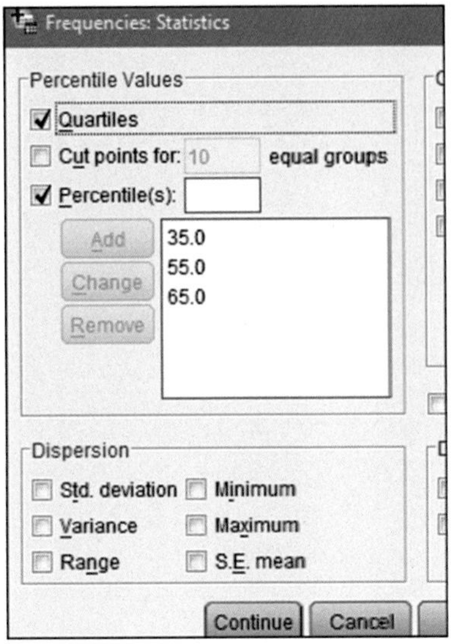

Fig. 4.16 Percentiles added

Fig. 4.17 Result table generated after the last step

Dialogs option. Choose the box plot option from the drop-down window that follows (Figs. 4.18 and 4.19).

Select the "Simple" box plot and the "Summaries of separate variables" option. Then click "Define". After that, add the required variable to the list. We will demonstrate the process using the scores for the physics test. Click OK, and the box plot should appear in a new window (Fig. 4.20).

Graphs	Utilities	Add-ons	Window	Help		

▮ Chart Builder...
▮ Graphboard Template Chooser...

Legacy Dialogs ▸

hysics	Math	VAR0000 01	VAR0000		VA

- Bar...
- 3-D Bar...
- Line...
- Area...
- Pie...
- High-Low...
- Boxplot...
- Error Bar...
- Population Pyramid...
- Scatter/Dot...
- Histogram...

39	54	.
68	65	.
100	100	.
75	79	.
88	93	.
82	97	.
56	59	.
81	94	.
74	71	.
91	94	

Fig. 4.18 Generating box plot

Fig. 4.19 Selecting required options for box plot

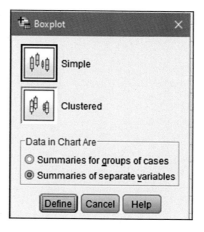

4.2.4 *Frequency Distribution*

Go to the "Analyze" toolbar as shown in Fig. 4.21. Look for the "Descriptive Statistics" option from the drop-down menu that appears. Select "Frequencies". A new window called "Frequencies" will pop up. Select the desired variable or variables. In this case, we will work with the "Score in Mathematics" variable, as depicted in Fig. 4.22. Once the desired variable is selected, click on the arrow in the middle to add it to the variable list (Figs. 4.23 and 4.24).

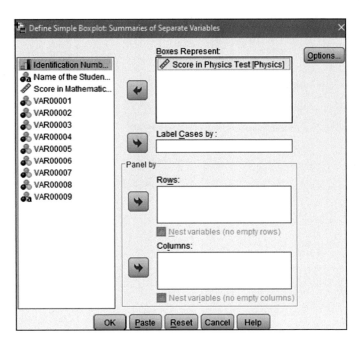

Fig. 4.20 Adding required variable for plot

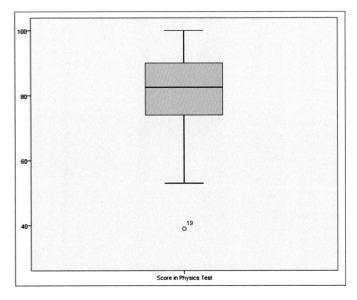

Fig. 4.21 Box plot generated

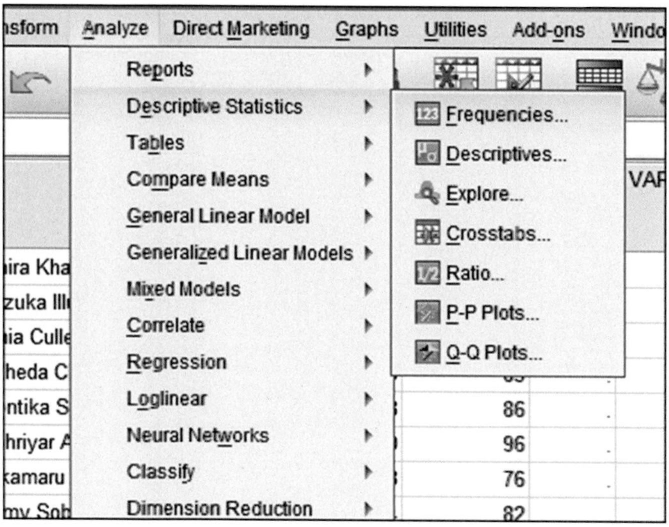

Fig. 4.22 Selecting frequencies

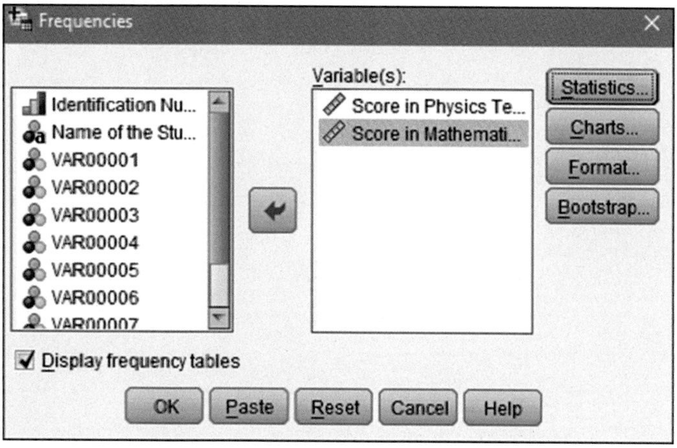

Fig. 4.23 Adding variables (steps)

4.2.5 Homework Problem

1. Dr Gilbert, a dietitian, is surveying his patients' BMI to check fitness trends. The following table shows the BMI of the patients. Calculate the following mentioned parameters to help him find the trend.

Score in Mathematics Test

		Frequency	Percent	Valid Percent	Cumulative Percent
Valid	50	1	3.3	3.3	3.3
	54	1	3.3	3.3	6.7
	59	1	3.3	3.3	10.0
	65	1	3.3	3.3	13.3
	71	1	3.3	3.3	16.7
	72	1	3.3	3.3	20.0
	76	1	3.3	3.3	23.3
	79	1	3.3	3.3	26.7
	80	1	3.3	3.3	30.0
	82	1	3.3	3.3	33.3
	83	1	3.3	3.3	36.7
	86	3	10.0	10.0	46.7
	87	2	6.7	6.7	53.3
	88	1	3.3	3.3	56.7
	89	2	6.7	6.7	63.3
	93	1	3.3	3.3	66.7
	94	2	6.7	6.7	73.3
	95	1	3.3	3.3	76.7
	96	1	3.3	3.3	80.0
	97	2	6.7	6.7	86.7
	98	1	3.3	3.3	90.0
	100	3	10.0	10.0	100.0
	Total	30	100.0	100.0	

Fig. 4.24 Frequency table for mathematics scores

Patient name	BMI
Clara	16.2
Twinkle	22.1
George	23.9
Robert	18.8
Catherine	14.4
Zara	25.2
Phillips	28.6
Andrew	27.3
Kayla	37.6
Liana	34.5
Russell	28.6

(continued)

(continued)

Patient name	BMI
Tanvir	20.0
Mustafy	32.5
Bella	29.5

(a)   Find the mean, mode, and median of the BMI of the patients.
(b)   Calculate the standard deviation and variance.
(c)   Find the 1st and 3rd quartile and 23rd, 47th, and 79th percentile.
(d)   Draw a box plot for this dataset.

4.3   Correlation

4.3.1   Scatter Plots

Plotting scatter graphs in SPSS is quite simple. Go to the Graphs toolbar and select the Chart Builder option from the drop-down menu that appears. A new window will pop up, as shown in Fig. 4.25. Click OK to continue to the Chart Builder window.

Suppose we want to see if the score in Mathematics is related to the score in Physics. Select the "Scatter/Dot" option in the Chart Builder window. Then drag the first template, "Simple Scatter", to the chart preview area as shown in Fig. 4.26. Next, drag the required variables to the X- and Y-axes (Figs. 4.27, 4.28 and 4.29).

4.3.2   Pearson's Correlation Coefficient

Let us find the correlation coefficient between the Mathematics and Physics scores. Go to the Analyze toolbar. From the drop-down menus that appear, select Correlation

Fig. 4.25 Opening chart builder

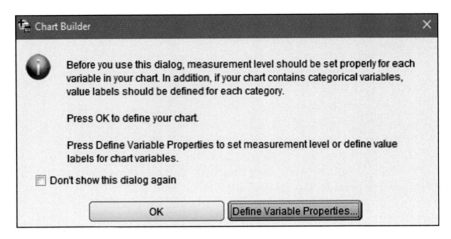

Fig. 4.26 Chart builder

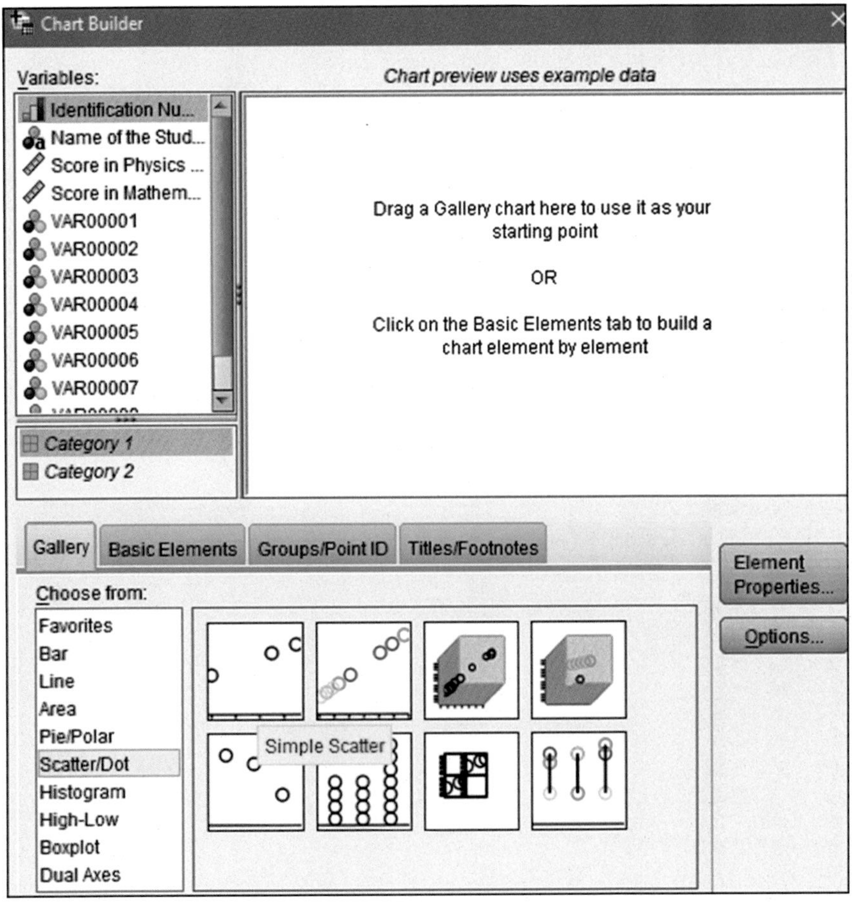

Fig. 4.27 Adding scatter template to preview area

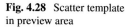

Fig. 4.28 Scatter template in preview area

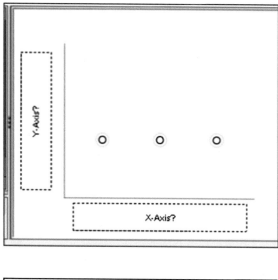

Fig. 4.29 Scatter plot

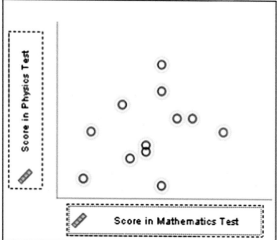

and then Bivariate. Add the desired variables to the variable list. Select the correlation coefficient required. We will be using Pearson's correlation coefficient here. Click OK, and the result should appear in a new window (Figs. 4.30, 4.31 and 4.32).

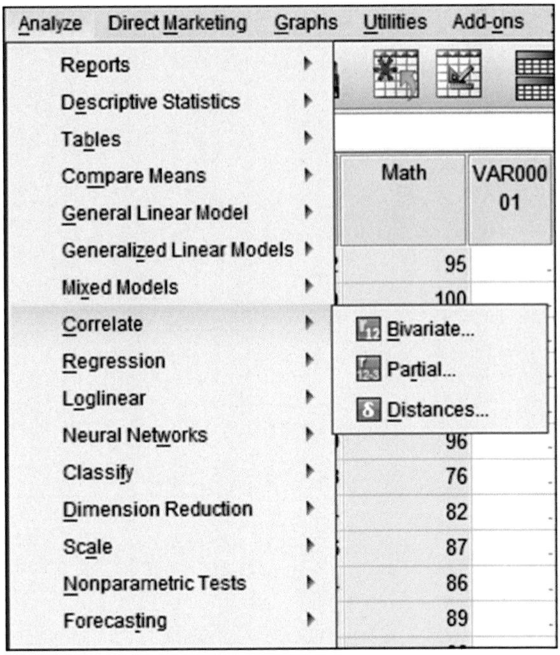

Fig. 4.30 Adding template to preview area

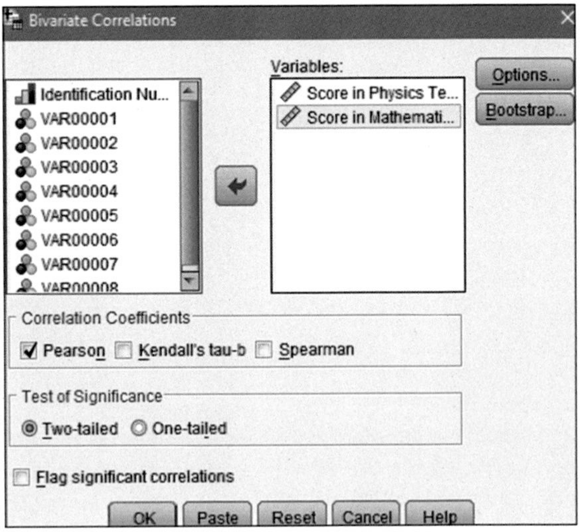

Fig. 4.31 Adding variables for scatter plots

→ Correlations

[DataSet0] C:\Users\PCFIX\Desktop\Mahira\Data Analysis Book\SPSS\Untitledl.sav

Correlations

		Score in Physics Test	Score in Mathematics Test
Score in Physics Test	Pearson Correlation	1	.884
	Sig. (2-tailed)		.000
	N	30	30
Score in Mathematics Test	Pearson Correlation	.884	1
	Sig. (2-tailed)	.000	
	N	30	30

Fig. 4.32 Correlation results generated

4.3.3   Homework Problem

1. An agricultural research organization tested to see if using a particular fertilizer increases food supply production.

Fertilizer(lbs)	Production of beans
2	4
1	2
3	4
2	3
4	6
5	5
3	5
7	7

(a)   Calculate the coefficient of correlation
(b)   Draw the scatter plot.

4.4   Regression Analysis

4.4.1   Simple Linear Regression

For regression analysis, go to the Analyze toolbar. From the drop-down menus that appear, select Regression. Then choose the desired regression model. We will explore the linear models here. Once the model type is selected, a new window should appear. Add the desired variables to the variable list (Fig. 4.33).

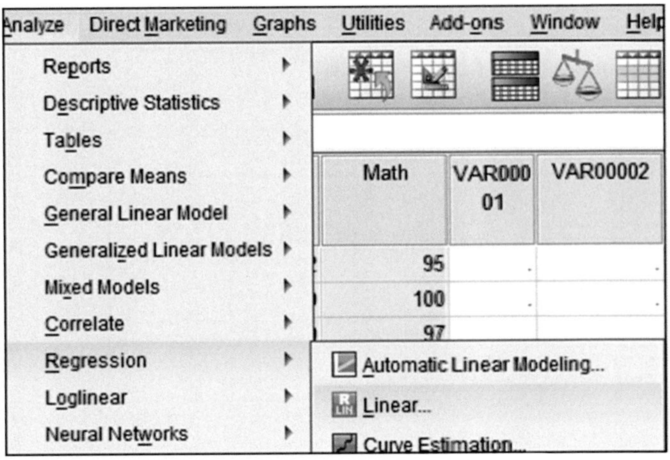

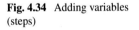

Fig. 4.33 Linear regression

Add the dependent and independent variables that are required. Since we want to study the effect of Mathematics on Physics, Mathematics is our independent variable, and Physics is the dependent one. Modify other properties as needed. Click OK, and the result should appear in a new window (Fig. 4.34).

Fig. 4.34 Adding variables (steps)

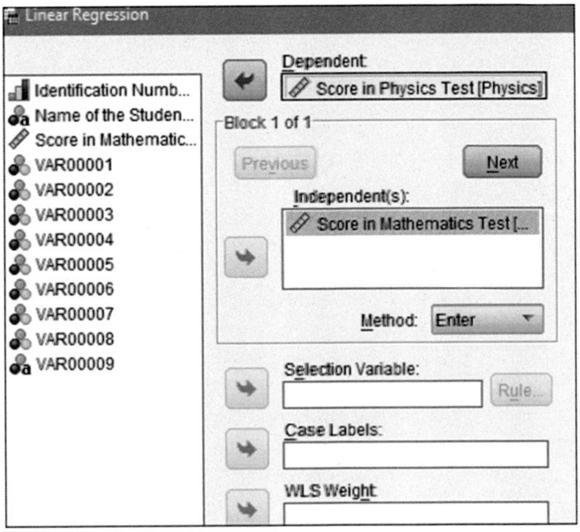

4.4.2 Homework Problems

1. The sales of tea and sugar at Sinha Departmental Store have varied this year due to the fluctuating prices of sugar and tea. The following table shows the trend.

Price of sugar	Price of tea	Sales
55	61	380
59	57	385
57	52	377
65	47	390
61	50	410
60	54	403
56	51	382
63	53	387

(a) Show with a regression model how the price of sugar affects store sales.
(b) Show with a regression plot how the price of both sugar and tea affect store sales.

4.5 Graphs and Charts

4.5.1 Line Graphs

Go to the Graphs toolbar and select the Chart Builder option from the drop-down menu that appears. A new window will pop up, as shown in Fig. 4.35. In the Chart Builder window, choose the Line graph option. Then drag the required template to the preview area. Add the desired variables to the X- and Y-axes (Figs. 4.36 and 4.37).

4.5.2 Coding the Data

The following table shows a dataset for the trophic status of lakes around Dhaka City. We will be using this dataset for the following few sections. The table displays four possible trophic states measured along different points of four lakes in Dhaka City.

Regression

[DataSet1] C:\Users\PCFIX\Desktop\Mahira\Data Analysis Book\SPSS\Untitled1.sav

Variables Entered/Removed[a]

Model	Variables Entered	Variables Removed	Method
1	Score in Mathematics Test[b]	.	Enter

a. Dependent Variable: Score in Physics Test
b. All requested variables entered.

Model Summary

Model	R	R Square	Adjusted R Square	Std. Error of the Estimate
1	.884[a]	.781	.773	6.737

a. Predictors: (Constant), Score in Mathematics Test

ANOVA[a]

Model		Sum of Squares	df	Mean Square	F	Sig.
1	Regression	4532.108	1	4532.108	99.866	.000[b]
	Residual	1270.692	28	45.382		
	Total	5802.800	29			

a. Dependent Variable: Score in Physics Test
b. Predictors: (Constant), Score in Mathematics Test

Coefficients[a]

Model		Unstandardized Coefficients B	Std. Error	Standardized Coefficients Beta	t	Sig.
1	(Constant)	2.473	7.835		.316	.755
	Score in Mathematics Test	.916	.092	.884	9.993	.000

Fig. 4.35 Regression results

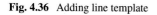

Fig. 4.36 Adding line template

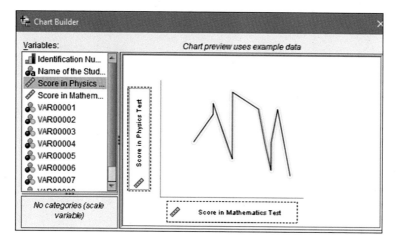

Fig. 4.37 Line graph

Sample	Trophic state	Sample	Trophic state
Banani		*Uttara*	
B1	Mesotrophic	U1	Oligotrophic
B2	Mesotrophic	U2	Mesotrophic
B3	Oligotrophic	U3	Oligotrophic
B4	Mesotrophic	U4	Mesotrophic
B5	Oligotrophic	U5	Mesotrophic
B6	Eutrophic	U6	Mesotrophic
B7	Eutrophic	U7	Eutrophic
B8	Mesotrophic	U8	Eutrophic
B9	Mesotrophic	U9	Eutrophic
B10	Oligotrophic	U10	Oligotrophic
Gulshan		*Hatirjheel*	
G1	Eutrophic	H1	Hypereutrophic
G2	Eutrophic	H2	Hypereutrophic
G3	Hypereutrophic	H3	Hypereutrophic
G4	Eutrophic	H4	Hypereutrophic
G5	Hypereutrophic	H5	Hypereutrophic
G6	Hypereutrophic	H6	Eutrophic
G7	Hypereutrophic	H7	Hypereutrophic
G8	Hypereutrophic	H8	Eutrophic
G9	Eutrophic	H9	Eutrophic
G10	Mesotrophic	H10	Eutrophic

There are two variables in this dataset—the location of sample collection and the trophic status at each point. Add both as string-type variables with nominal measure type in the Variable View. No missing values need to be defined, and the standard column width of 8 should suffice.

Once the variables have been defined, go to the Data View and enter the data, as shown in Fig. 4.38. Notice how the terms used to define the trophic states are not simple. Manually typing these would be tedious and time-consuming. Now say we want to encode the trophic state names, considering that there are four distinct categories (Fig. 4.39).

Fig. 4.38 Defining variables (steps)

	Sample	State			Sample	State
1	B1	Mesotrophic		20	G10	Mesotrophic
2	B2	Mesotrophic		21	U1	Oligotrophic
3	B3	Oligotrophic		22	U2	Mesotrophic
4	B4	Mesotrophic		23	U3	Oligotrophic
5	B5	Oligotrophic		24	U4	Mesotrophic
6	B6	Eutrophic		25	U5	Mesotrophic
7	B7	Eutrophic		26	U6	Mesotrophic
8	B8	Mesotrophic		27	U7	Eutrophic
9	B9	Mesotrophic		28	U8	Eutrophic
10	B10	Oligotrophic		29	U9	Eutrophic
11	G1	Eutrophic		30	U10	Oligotrophic
12	G2	Eutrophic		31	H1	Hypereutrophic
13	G3	Hypereutrophic		32	H2	Hypereutrophic
14	G4	Eutrophic		33	H3	Hypereutrophic
15	G5	Hypereutrophic		34	H4	Hypereutrophic
16	G6	Hypereutrophic		35	H5	Hypereutrophic
17	G7	Hypereutrophic		36	H6	Eutrophic
18	G8	Hypereutrophic		37	H7	Hypereutrophic
19	G9	Eutrophic		38	H8	Eutrophic
20	G10	Mesotrophic		39	H9	Eutrophic
				40	H10	Eutrophic

Fig. 4.39 Data added

Name	Type	Width	Decimals	Label	Values	Missing	Columns
Sample	String	3	0	Lake Sample Collection Point	None	None	8
State	String	20	0	Trophic Status of Point Along Lake	None	None	12

Fig. 4.40 Modifying values

Go to the Variable View to edit the "Values" property. Click on the three dots beside "None". This will open up the "Value Labels" window. Write the desired result (each trophic state's name) beside the "Label" option. Insert a number next to the "value" option to denote the result. Since we have four possible results, let us number them from 1 to 4. After completing one entry, click "Add" and input the remaining codes similarly (Figs. 4.40 and 4.41).

Once all the values are added, click OK and return to the Data View. We can now type the number 2, for example, and click on the "Value Labels" option to convert it to the word Mesotrophic and vice versa. This process can be repeated for all the results. Thus, we only need to type numbers from 1 to 4, and SPSS will automatically correspond the numbers to our coded words (Fig. 4.42).

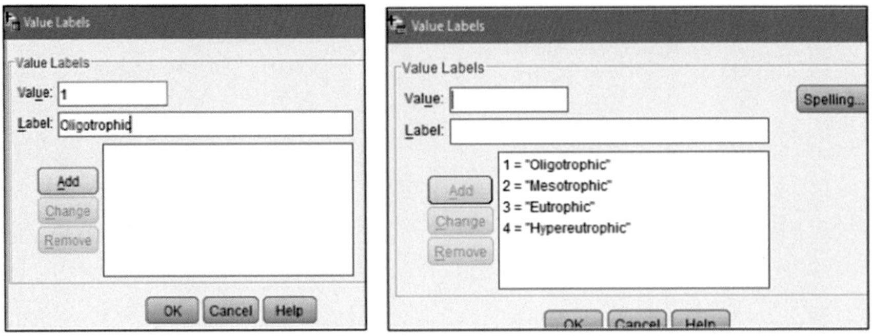

Fig. 4.41 Coding values

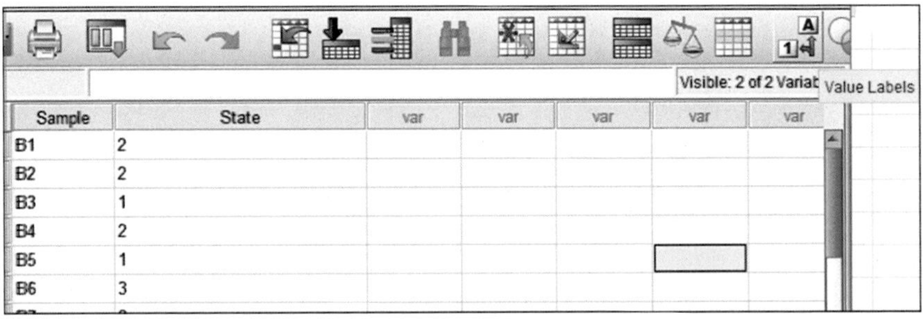

Fig. 4.42 Coded values

4.5.3   Bar Charts

Once the data entry is completed, it is pretty simple to represent the dataset in the form of a bar chart. As shown in Fig. 4.43, click on the Graphs toolbar. In the drop-down menu that appears, select Legacy Dialogs, and a second drop-down menu should appear. Select the Bar option from there. A pop-up window will appear. Select the "Summaries for groups of cases" option and click Define (Figs. 4.44 and 4.45).

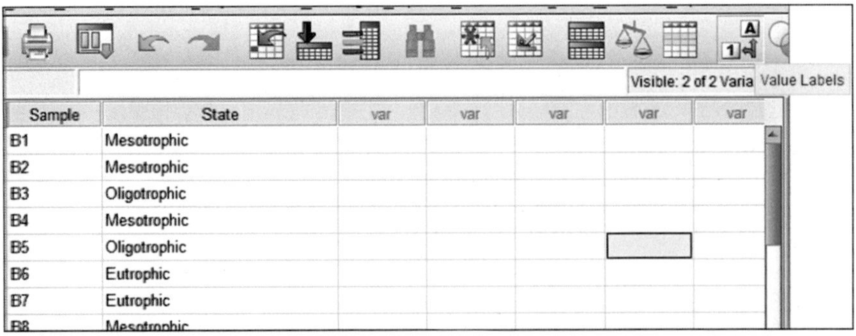

Fig. 4.43 Switching between codes and labels

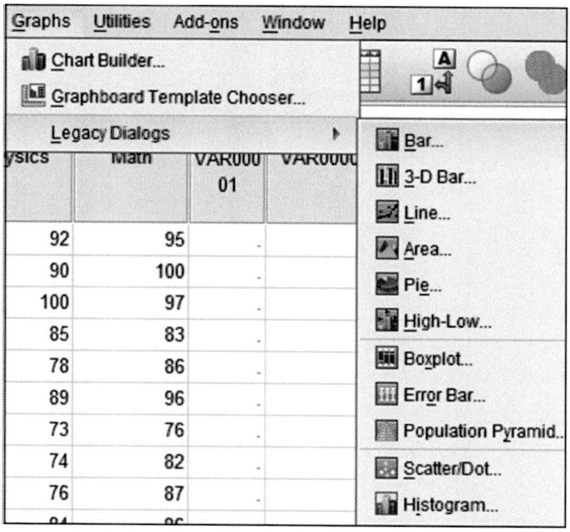

Fig. 4.44 Selecting bar chart

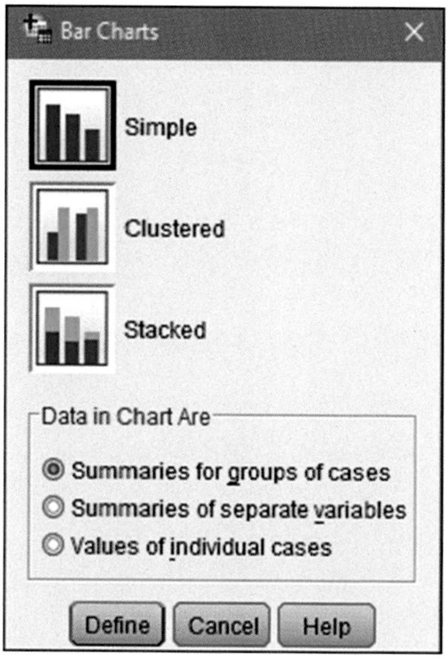

| Graphs | Utilities | Add-ons | Window | Help |

- Chart Builder...
- Graphboard Template Chooser...
- Legacy Dialogs ▶

| Bar... |
| 3-D Bar... |
| Line... |
| Area... |
| Pie... |
| High-Low... |
| Boxplot... |
| Error Bar... |
| Population Pyramid.. |
| Scatter/Dot... |
| Histogram... |

ysics	Math	VAR000 01	VAR0000
92	95	.	
90	100	.	
100	97	.	
85	83	.	
78	86	.	
89	96	.	
73	76	.	
74	82	.	
76	87	.	

Fig. 4.45 Bar chart options

Bar Charts

- Simple
- Clustered
- Stacked

Data in Chart Are
- ● Summaries for groups of cases
- ○ Summaries of separate variables
- ○ Values of individual cases

Define | Cancel | Help

Add the "Trophic State of Points Along Lake" to the Category Axis and click OK. Once the bar chart appears in a new window, right-click on it and select "Edit Content" and "Separate Window" (Figs. 4.46 and 4.47).

Click on the Elements option. A drop-down menu will appear. Choose "Data Label Mode" from there. The cursor should change into a black icon. Click on each bar to label it (Fig. 4.48).

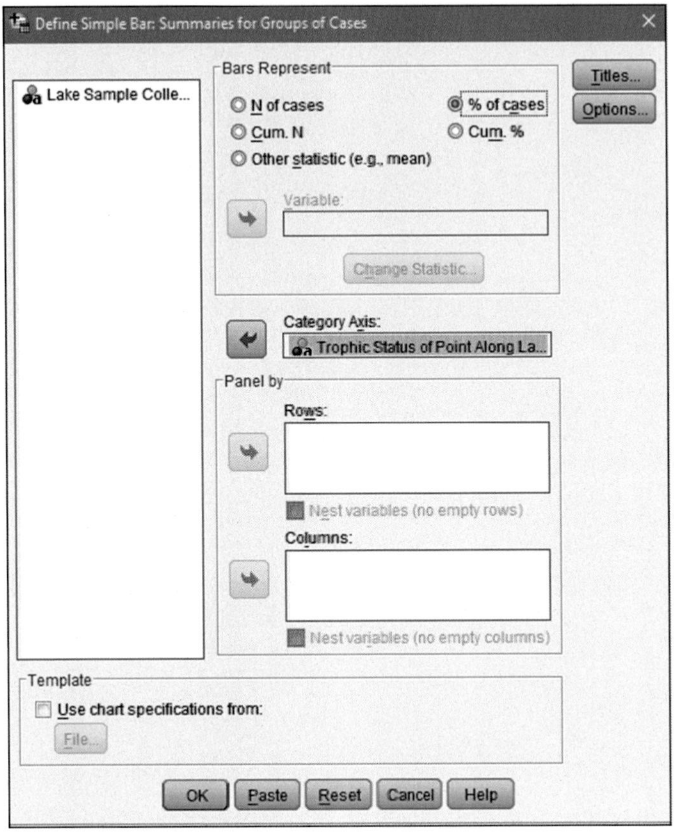

Fig. 4.46 Adding variables to category axis

4.5.4 Pie Charts

Once the data entry is completed, it is pretty simple to represent the dataset in the form of a pie chart. But first, let us generate a frequency table. As shown in Fig. 4.49, go to the Analyze toolbar. A drop-down menu will appear. Click on the descriptive statistics option. A second drop-down menu will appear. Select the Frequencies option from there. Add the desired variable to the variable list and click OK. The frequency table will appear in a new window, as illustrated in Fig. 4.50.

Now that the frequency table has been generated, we can see the percentage of each class present in the dataset. Follow the upcoming procedure for drawing the pie chart.

As shown in Fig. 4.51, click on the Graphs toolbar. In the drop-down menu that appears, select Legacy Dialogs, and a second drop-down menu should appear. Select the pie option from there. A pop-up window will appear. Select the "Summaries for groups of cases" option and click Define (Figs. 4.52 and 4.53).

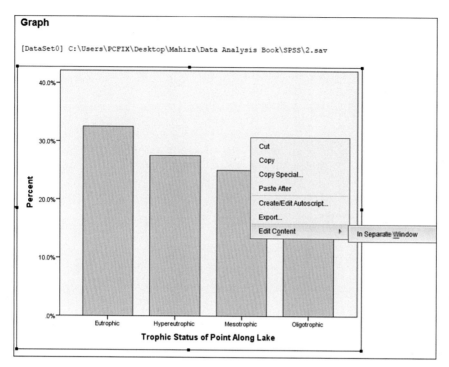

Fig. 4.47 Editing bar chart

Once the pie chart appears, we can edit it to add labels on each segment. Doing so will enable us to see the percentage of each class present in the dataset while also visualizing the proportions. Right-click on the pie chart and select "Edit Content", as depicted in Fig. 4.54. Then choose "Separate Window". Click on the Elements option as shown in Fig. 4.55. A drop-down menu will appear. Choose "Data Label Mode" from there. The cursor should change into a black icon. Click on each segment to label it (Fig. 4.56).

4.5.5 Histogram

Figure 4.57 shows the weight of different students in a class, measured in kilograms. This dataset can be visually represented using a histogram. SPSS offers a simple process for plotting histograms. First, define the variable and modify all the required properties (Fig. 4.58).

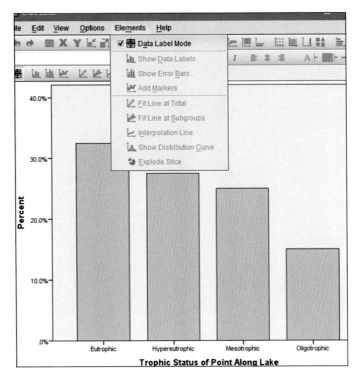

Fig. 4.48 Adding data labels

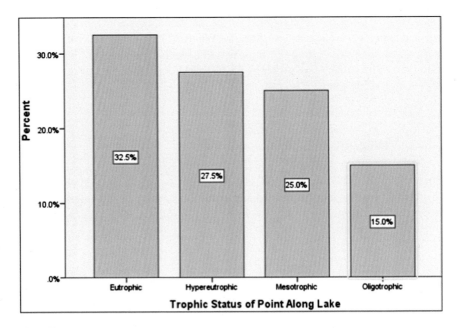

Fig. 4.49 Bar chart

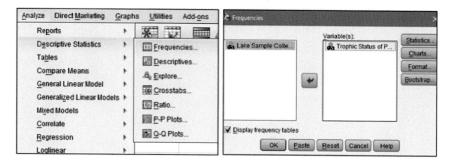

Fig. 4.50 Generating frequency table

Frequencies

[DataSet0] C:\Users\PCFIX\Desktop\Mahira\Data Analysis Book\

Statistics

Trophic Status of Point Along
Lake

N	Valid	40
	Missing	0

Trophic Status of Point Along Lake

		Frequency	Percent	Valid Percent	Cumulative Percent
Valid	Eutrophic	13	32.5	32.5	32.5
	Hypereutrophic	11	27.5	27.5	60.0
	Mesotrophic	10	25.0	25.0	85.0
	Oligotrophic	6	15.0	15.0	100.0
	Total	40	100.0	100.0	

Fig. 4.51 Frequency results

Fig. 4.52 Selecting pie chart

Graphs	Utilities	Add-ons	Window	Help
Chart Builder...				A
Graphboard Template Chooser...				1
Legacy Dialogs			▶	Bar...
var	var	var		3-D Bar...
				Line...
				Area...
				Pie...

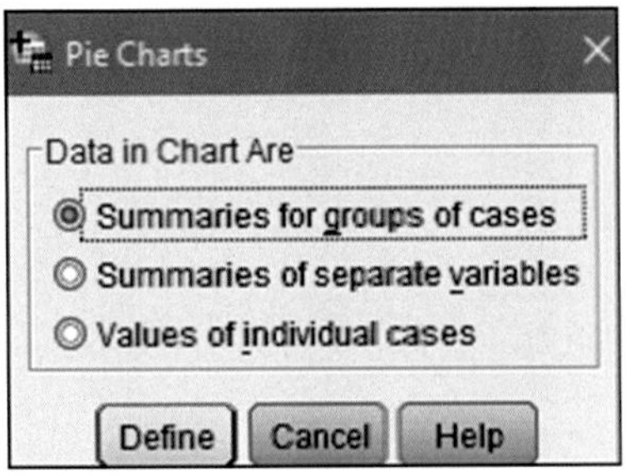

Fig. 4.53 Options for pie chart

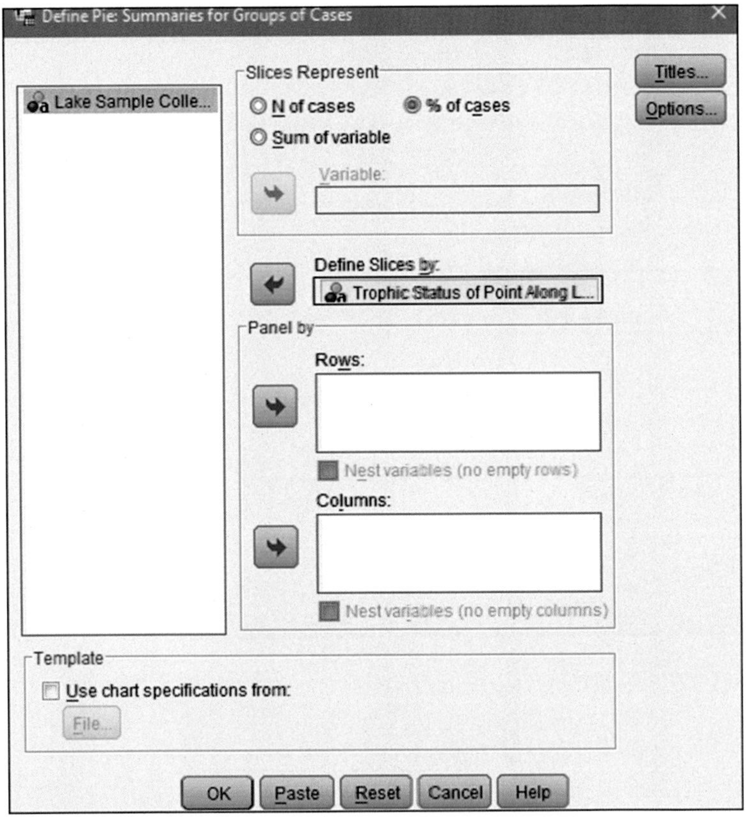

Fig. 4.54 Adding variables for pie charts

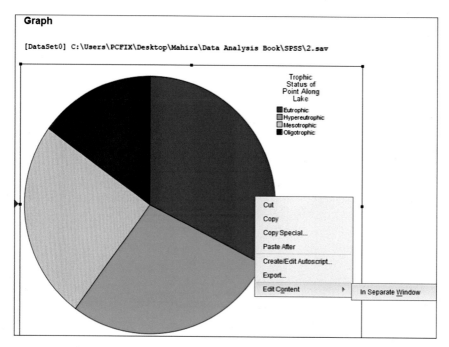

Fig. 4.55   Modifying pie chart

   The name and description of the variable can be anything. There is no fixed rule for labeling the variables. Select the numeric type and choose a suitable value for the width of the data. There is no missing value to focus on in this example, and the data does not need to be coded. The Scale Measure should be used since it is a continuous variable (Fig. 4.59).

   Once the variable is defined, go to the Data View. Enter the weight values under the defined variable column. As shown in Fig. 4.60, click on the Graphs toolbar. In the drop-down menu that appears, select Legacy Dialogs, and a second drop-down menu should appear. Select the Histogram option from there (Fig. 4.61).

   In the pop-up window that appears, select the "Weight of the Students (Kg)" (desired variable). Click on the arrow in the middle to add it to the variable list. Click OK, and the histogram should appear in a new window, as shown in Fig. 4.62.

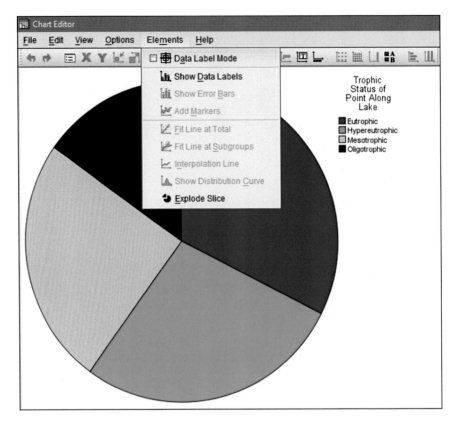

Fig. 4.56 Adding data labels for pie chart

4.5.6 Homework Problems

1. In a gathering, the attendees were asked their favorite drink. The following table shows the summary of the survey. Draw a bar chart diagram showcasing the various favorite drinks of the attendees.

Favorite drink	No. of attendees
Milk	7
Soda	15
Water	9
Juice	12
Tea	17

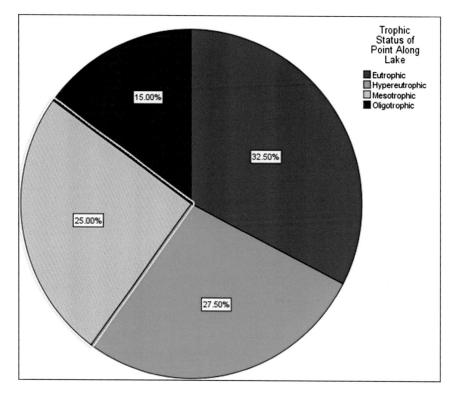

Fig. 4.57 Pie chart

Fig. 4.58 Weight of
students in kg

A
Weight of Students (Kg)
53
62
68
55
59
65
67
72
79
77
75
65
63
52

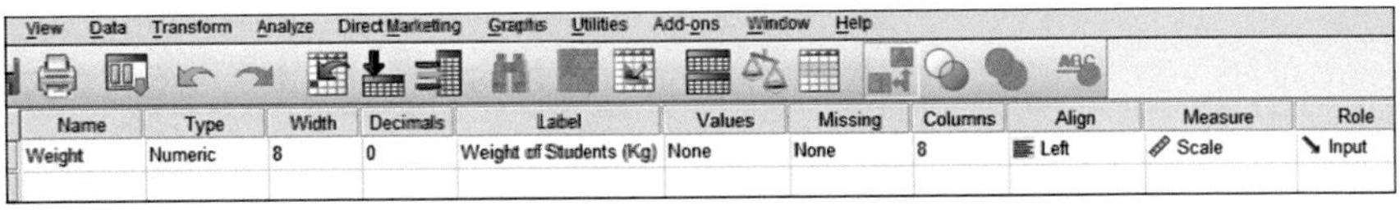

Fig. 4.59 Defining variables for the weight of students

	Weight
1	53
2	62
3	68
4	55
5	59
6	65
7	67
8	72
9	79
10	77
11	75
12	65
13	63

Fig. 4.60 Data entry

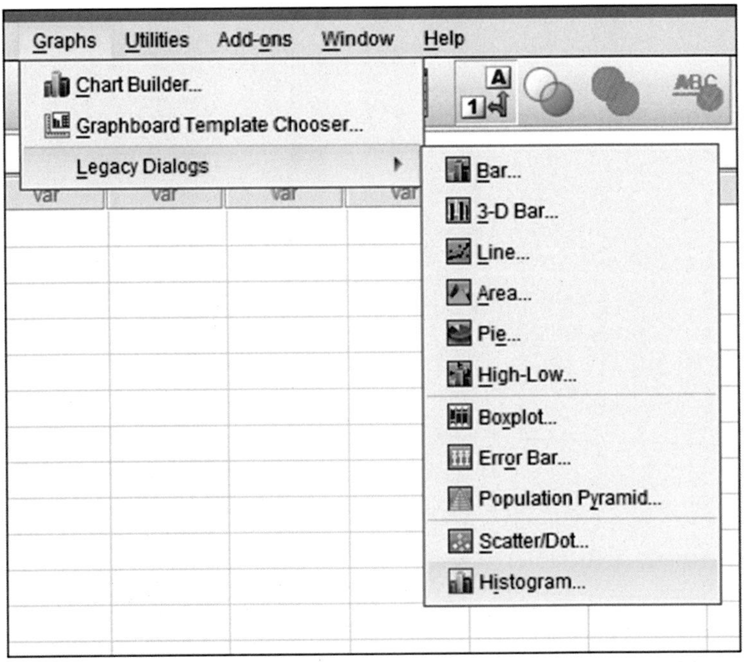

Fig. 4.61 Choosing histogram

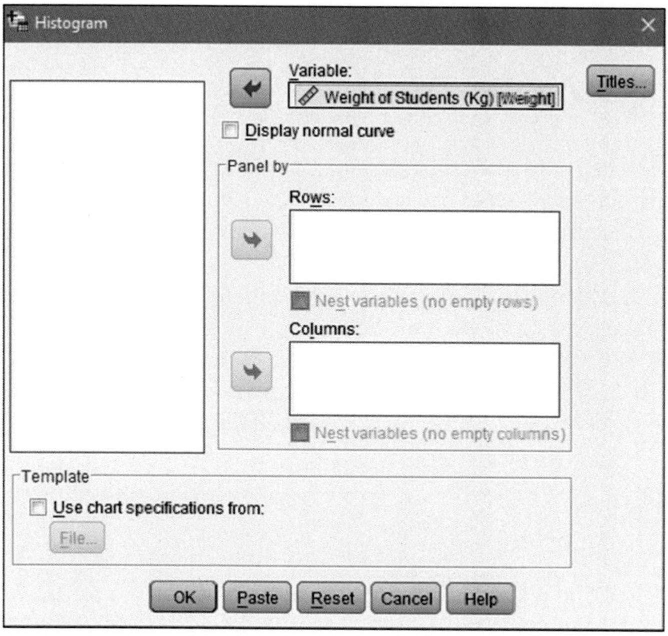

Fig. 4.62 Adding variable

2. Class 9 of Saint Lawrence High School students were asked for their favorite flavor of ice cream. The following table shows their preference. Prepare a pie chart of the dataset.

Favorite flavor	No. of students
Chocolate	7
Vanilla	15
Strawberry	9
Black forest	12
Butter scotch	17

4.6   Inferential Statistics

4.6.1   Analysis of Variance (ANOVA)

Analysis of variance, ANOVA, is an essential part of data analysis, and SPSS provides a straightforward approach to performing this analysis. Figure 4.63 shows a dataset comprising students' test scores from three different sections, Section A, Section B, and Section C (Fig. 4.64).

We want to determine if there is a difference in the performance between students of different sections. First, we must define our null and alternative hypotheses for an ANOVA test. Let the null hypothesis be that there is no difference between the scores of the three sections. The alternative hypothesis is that there is a difference between at least two sections; i.e., they are not all the same (Fig. 4.65).

Fig. 4.63 Histogram

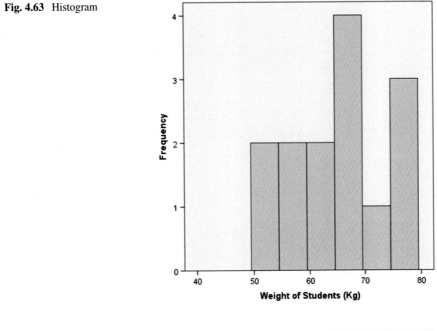

Fig. 4.64 Test scores in different sections

A	B	C
49	42	40
50	40	42
45	41	38
47	43	44
48	39	41
46	40	42

Name	Type	Width	Decimals	Label	Values	Missing	Columns	Align	Measure	Role
Scores	Numeric	8	0	Test Scores	None	None	8	▦ Right	⬕ Scale	↘ Input
Section	Numeric	8	0	Section Name	{1, A}...	None	8	▦ Right	⬗ Nominal	↘ Input

Fig. 4.65 Defining variables

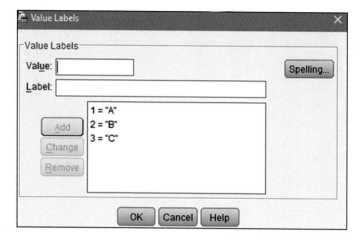

Fig. 4.66 Adding value labels

Define the variables as shown above. The type should be numeric for both variables. Add the variable description under the Label option. There are no missing values here that we need to consider particularly. The default column width of 8 should be sufficient, but we can always manually expand or contract the column width in the Data View as needed. Select the Scale Measure for the "Scores" variable and Nominal Measure for the "Section" variable

Select the three dots under the Values property. Type in the desired name/label next to the Label option and the corresponding code next to the Value option. We are coding the three section names A, B, and C as 1, 2 and 3, respectively. The coding for such short labels is not necessary or even particularly useful. However, if the labels were longer, it would save time and make the data entry process much less tedious (Figs. 4.66 and 4.67).

Go to the Data View and type in all the codes, i.e., 1, 2, and 3. Next, click on the Value Labels option on the top toolbar. This tool switches the codes to their respective labels, as we have already seen in an earlier section. Once switched, the entries should be displayed as "A", "B", and "C". If needed, one can always switch back to the coded form.

After the data entry is complete, click on the Analyze toolbar. Select the Compare Means option from the drop-down menu. Then select the One-Way ANOVA option as shown in Fig. 4.68. Select the desired variable to perform the ANOVA test. Add the variables to the dependent and independent lists. In this example, the dependent variable is the test score, while the section is the independent variable. Once the

Fig. 4.67 Coded data

Scores	Section	var
49	1	
50	1	
45	1	
47	1	
48	1	
46	1	
42	2	
40	2	
41	2	
43	2	
39	2	
40	2	
40	3	
42	3	
38	3	
44	3	
41	3	
42	3	

Fig. 4.68 Code switched to labels

Scores	Section	var
49	A	
50	A	
45	A	
47	A	
48	A	
46	A	
42	B	
40	B	
41	B	
43	B	
39	B	
40	B	
40	C	
42	C	
38	C	
44	C	
41	C	
42	C	

variables are added, click OK, and the results should appear in a new window as usual (Figs. 4.69, 4.70 and 4.71).

Now, observe the results and state the inference. Since the significance value is less than 0.05, we can say that the null hypothesis is rejected. The results indicate

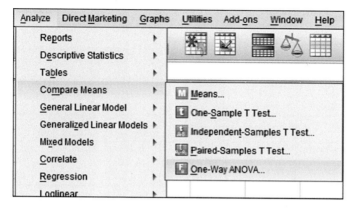

Fig. 4.69 Selecting ANOVA

Fig. 4.70 Adding variables for ANOVA test

an inconsistency between the test scores obtained by students in different sections. There could be several different reasons for this. However, the ANOVA test does not give us any justification for its results. We can speculate that the teaching methods or grading system might have differed between the classes. Alternatively, other external factors could affect the average performance in the different classes.

Oneway

[DataSet1] C:\Users\PCFIX\Desktop\Mahira\Data Analysis Book\SPSS\4.sav

ANOVA

Test Scores

	Sum of Squares	df	Mean Square	F	Sig.
Between Groups	169.333	2	84.667	25.831	.000
Within Groups	49.167	15	3.278		
Total	218.500	17			

Fig. 4.71 ANOVA results

4.6.2   Homework Problem

1. The following dataset represents the number of lemonades sold by three teams at a school festival. Is there any significant difference between the sales? Perform an ANOVA test to justify the answer. What could be the possible reasons for any discrepancy?

Green team	Blue team	Red team
35	41	31
36	42	32
37	43	34
34	45	37
39	40	36

4.6.3   Exercises

1. The sales head of Bay Watch was asked to summarize the watch sales of that year. The following table shows the number of sales each month.

Month	Sales
January	79
February	85

(continued)

(continued)

Month	Sales
March	68
April	89
May	94
June	88
July	72
August	80
September	91
October	73
November	79
December	99

(a) Calculate the mean, mode, median, and standard deviation.
(b) Draw a box plot of the sales.

2. The following dataset represents the number of items received by three groups at a school club as part of the winter donation program. Is there any significant difference between the donations received? Perform an ANOVA test to justify the answer. What could be the possible reasons for any discrepancy?

Group A	Group B	Group C
35	41	31
36	42	32
37	43	34
34	45	37
39	40	36

3. In a traffic survey, the percentages of modal share were obtained. Draw a pie chart representing the results.

Vehicle	Percentage
Car	25
HGV	15
Motor bike	30
Bicycle	12
Bus	18

4. Rahim is a college student. The following table shows the test scores he obtained in the previous semester. Prepare a bar chart of the dataset.

Subject	Score
Mathematics	87
History	76
English	85
Physics	92
Chemistry	98